W9-BVI-718

Every wave, every ebb wears away relentlessly those things that man has planted in the sea's domain . . .

FOG AND SUN
SEA AND STONE

THE MONTEREY COAST

. . . or that nature has set to mark its boundaries.

FOG AND SUN
SEA AND STONE

THE MONTEREY COAST

PHOTOGRAPHS AND WORDS BY STEVE CROUCH

EZ Nature Books
San Luis Obispo, California

For Matthew Walter

Man's works that long survive the sea take on the patina of time, the dignity of age.

ISBN 0-945092-25-3(Hard Cover), 0-945092-26-1(Soft Cover)

Library of Congress Catalog Number 80-66365

© MCMLXXX by Graphic Arts Center Publishing Company
Photographs © MCMLXXX by Steve Crouch

Publisher MCMXCI Revised Edition - EZ Nature Books
P.O. Box 4206, San Luis Obispo, CA 93403

This book authorized and published under special arrangement with
Graphic Arts Center Publishing Company in Portland, Oregon.

Printed in the United States of America

CONTENTS

PROLOGUE

Railroad stations are strange, inhospitable places in which to find one's self even in the best of times. During the war years of the forties, they were grim way stations along the route into an uncertain and ominous future. The Salinas depot was neither better nor worse than most. It was a depressing introduction to what was to become home to me for more than half my life.

Salinas lies only a few miles inland from the great western sea, almost at the end of the long river valley that bears its name, through which the Southern Pacific makes its way between Los Angeles and San Francisco. A scant dozen miles away, along the shore of the Bay of Monterey, was Fort Ord — in those days the last stop before the fearful Adventure that waited on some far shore. It was there I had been sent by the army to await shipment to who-knew-where. Until that time arrived, I planned to live in one of the towns on the nearby Monterey Peninsula.

In summer cold winds blow in from the Pacific and bring chill fogs up the valley. As I stood with my small family shivering in the thin sunshine of that July afternoon, a strong breeze cut through my cotton summer uniform and left me longing for the hot lands so recently left behind halfway across the continent.

The bus bound for Monterey pulled away from the railroad station and made its way through the streets of Salinas, out past the sprawl of produce sheds and packing plants. Leaving the town behind, it went through flat miles of carefully tended lettuce fields, crossed the sluggish Salinas River on a narrow, rattling steel bridge, then began its climb into low foothills skirting Mount Toro, northern bastion of the Santa Lucia Range.

Along the dry stream courses were enormous sycamore trees and grazing flocks of sheep guarded by swift dogs and watchful shepherds — Basques from Spain, I later learned. On the slopes grew oaks of a kind new to me: great, gnarled giants whose twisted limbs often turned back almost into the earth. They clustered thick on the hillsides, and no underbrush grew beneath them, so the land resembled a well-kept park.

The bus bumped along at a moderate clip, the highway in those days being somewhat less than it is now. There was time to savor the changes brought by each turn in the road, each crest surmounted, each valley crossed. The grasses were dry, for the rains were long since past, and groves of oaks made great splotches of gray-green against the golden hills.

The gilded land turned slowly to warm gray as the sun was shut out by fog creeping in from the sea. Long streamers of Spanish moss began to appear, festooning the branches of the oaks and heralding a change in climate. Here and there were pines, growing more and more profusely until at last they became the dominant tree, and the road made its way through a thick forest of them.

The bus rolled on over the endless series of foothills that lay below the higher ridges and came at last to the outer limits of Monterey. There towering groves of blue Australian eucalyptus trees surrounded an emporium of food and drink whose red neon sign announced through the fog, "The Blue Ox." Just beyond lay the immaculate grounds of the venerable Hotel Del Monte, a caravanserai that had been patronized for more than sixty years by the wealthy and those who would appear so, but at the moment was serving as a training school for prospective naval aviators.

The way led down through rows of neat, small houses, each surrounded by white picket ramparts, where lived most of the Sicilian fisherfolk who manned the sardine fleet. The houses were close together as befit the gregarious nature of the people in them.

As in most towns with military establishments, the bus station sat in close proximity to a sprawl of short-order restaurants, raucous bars, down-at-the-heels hotels, pool halls, tattoo parlors, and pawn shops. Nearby was the waterfront and the harbor, which played host to more than half a hundred fishing boats. The streets were busy with soldiers and sailors, Italian and Portuguese fishermen in rubber boots and knitted caps, assorted floaters, and an occasional itinerant whore there to garner what she could from the largess of a military payday.

Down the street a block or two away lay Alvarado Street, the town's main thoroughfare, lined by one- and two-story buildings left over from the last century. Here and there a vintage adobe was wedged in among the wooden false fronts. The result was a mixture of Mexican village, New England town, and Hollywood movie-lot western cow town. It had an air about it all its own — like no other town I had ever seen.

On three sides of town loomed hills, thickly cloaked in pine and oak. On the fourth side was the bay, gray this late afternoon under an overhang of fog.

The road to Carmel, our ultimate destination for the day, lay athwart the shoulder of the mountain to the south. The Carmel bus, to which we had transferred, had but few riders when it pulled out of the terminal and began the long pull up Abrego Street to the bottom of the hill that marked the edge of Monterey. There the driver shifted to his lowest gear, and with its diesel engine protesting in a steady growl, the bus labored up the grade. Near the top we reached the level of the fog layer and suddenly moved into another world. The pines on each side of the road thinned to a few as the rest disappeared in the fog. Even those by the road's edge became only tall stumps, their tops hidden somewhere in the upper layers of the mist. The eye played tricks; dark, sinister highwaymen lying in wait in the fog suddenly became bushes as we passed, then reverted again to wraiths.

The headlights of the bus cast a weak glow on the fog ahead — less a guiding light for the driver than a warning for other ghostly machines passing in the opposite direction. Even the laboring of the engine seemed strangely muted by the fog. No one spoke during this gray passage.

Once upon the crest of the hill, the bus wandered through a small upland meadow, turned off into a side road to Carmel, and began slowly to descend. The fog thickened here on this slope facing the open sea. Through the trees small houses, each distinctly different from its neighbor, floated dimly into sight, then back into the murk. In some, lights were on even though it was not yet twilight; the fog softened the lighted windows into small pastel patches in the mist.

At length the bus ground to a halt in Carmel. The other passengers abandoned it and disappeared into the fog, leaving us standing alone and uncertain on the sidewalk surrounded by our luggage. The street was lined with small stores, whose windows were brighter than the daylight. Down its center stood a row of magnificent pines — pines so huge that they must have been as old as the street itself. And all about on every side were other trees — some close by, others towering beyond the buildings.

Suddenly I became aware of an unfamiliar sound, a mere background to the street noises at first, but now so insistent as not to be ignored — the crash of waves breaking upon the shore at the foot of the village.

Standing there in the fading light, wrapped in fog, and listening to the ocean's measured pulse, I realized that I had reached the place where I could be content to spend the rest of my days.

Forty years have come and gone — and nothing has changed my mind.

Salt winds and ocean fogs have bent, twisted, and weathered the cypress that stand tenaciously upon the granite seacliffs.

THE
ENDURING
COAST

*Cormorants like to perch on rocks just beyond
the pounding waves of winter storms.*

THE PENINSULA

The long sandy beach begins not far from Santa Cruz and forms a great crescent that only ends where the spine of the Monterey Peninsula comes down to meet the sea, under the hill where the Presidio of Monterey sits with its old polished cannon looking over the place where Sebastian Vizcaíno took possession of this land for Spain.

Except when storms or exceptional tides drive the sea up against the cliffs, you can stroll almost the whole length of the bay and never get your feet wet, except perhaps in crossing the Salinas and Pajaro rivers when winter rains have filled their banks and broken through the sandbars that block their way to the sea during most of the year. The entrance to Moss Landing — a break in the coast where Elkhorn Slough, vestigial remnant of a river of considerable size that no longer flows there — presents a barrier too deep to wade, but one that can be circumambulated.

There are wondrous things to be seen along this magnificent edge of the land. Near the north end there are stones lying in the sand with strange treasures locked within them — small animals that lived and died untold millions of years ago and left their skeletons here for treasure seekers of today to discover. Below the high cliffs north of the Pajaro River, thousands of sand dollars — round, flat echinoderms with a tracery of delicate fronds on their backs — are cast up by the sea and left stranded by the tides. Some play host to tiny barnacles cemented to their backs, a most accommodating relationship.

Near Moss Landing are the first of the low dunes that stretch almost to Monterey, forming a barrier between the sea and the fields of artichokes that cluster about the mouth of the Salinas River. Queer succulent plants grow on these dunes — pickleweed and sea rocket, which thrive where the air is heavy with salt. In the bunches of tough grasses that anchor the dunes in place, sparrows search for insects and fallen seed; perched atop bush lupines, their black-and-white striped heads turned into the wind, they inspect their world much as a proud landholder might seek out the highest point on his estate and survey the acres spread before him.

Long lines of brown pelicans fly back and forth across the beach, from the open sea to the marshes of Elkhorn Slough, like

The shoreline fairly shimmers when the sun,
low on the horizon, turns rocks and sand to gold.

formations of ancient, lumbering bombers. There are other sea and land birds, too, in great profusion. After great storms at sea, birds are encountered here far from their normal haunts. One recent storm blew before it a magnificent frigate bird, a thousand miles north of his usual range. He found shelter here until the storm subsided, then took to the air and for several days wheeled effortlessly above the slough and the harbor at Moss Landing. At length, finding nothing to his liking, he tilted his wings southward and disappeared, bound for the warm beaches of Baja California where the rest of his kind were congregated.

At another time, for three years in a row, a pair of flamingoes came from somewhere to summer in the marsh behind the dunes, exotic tourists probably on sabbatical from a zoo far to the south. They made a startling sight, standing on one leg there beside the highway as the late afternoon sun turned their pink feathers into scarlet flame.

These same mudflats and intertidal basins are a haven for uncounted numbers of shorebirds — some natives who live their whole lives there, some from the north who come to escape the rigors of winter. At times they are almost equaled in number by the flocks of birdwatchers who line the margins of the marshes with field glasses at the ready. Both groups eye each other — one avidly, the other warily; one utters cries of delight, the other cries of alarm. One group departs with new checkmarks in its bird books, the other watches the departure with relief.

Further south along the dunes, not far from the rifle ranges of Fort Ord, where the sounds of battle sometimes drown out the constant, restless sound of the sea, men suspended from great, colorful wings sail back and forth along the updrafts that occur when winds from the sea strike these sandy bluffs and rise up in continuous invisible waves, a phenomenon much sought after by hang gliders.

At the southern end of this long beach that began in distant haze on the far side of California's largest open bay, the hills come down to meet the sea to form a protected cove that has caught the eye of mariners since Vizcaíno first led his flotilla of Spanish galleons into it almost four hundred years ago. Here lies the town of Monterey.

On a wharf at the very end of the beach, there used to be a cannery, the first of its purpose in western North America. It was built to package the small sardines that once lived in great plentitude in the waters of this bay. It has long been gone, a victim of the ravages of the sea which tore at it and the greed of men who depleted the fish on which its existence was based. Now the cannery and its wharf are gone. The wharves that remain are the domain of tourists, who flock there to buy trifling souvenirs, and of astute shopkeepers who supply that eternal need. Yachts and pleasure boats fill the berths behind the Coast Guard's stone breakwater, where purse seiners used to ride at anchor.

The last rocks of the peninsula disappear into the sea west of the breakwater. Small circular beaches lie in coves at the base of sea bluffs. Rocky promontories project into the bay, forming islands against which the sea crashes in an explosion of spray.

The sea's wash brings up gifts from the deep to leave them upon the sand — a delight to the beachcomber's eye.

Thousands of cormorants and gulls and kittiwakes perch upon these rocks, and sea lions and seals haul out on them to sleep and to argue; their hoarse barking and yelping is a trademark of this coast.

On these rocky bluffs are the remains of the canneries that once lined Cannery Row. West of them sit the houses of Pacific Grove; some are almost as old as the last-built Mexican adobes in Monterey, others as new as yesterday. In the middle of Pacific Grove's waterfront is a small tree-filled park that juts into the bay. It is called Lover's Point, and even to most people who live here the name brings visions of young lovers trysting beneath the cypresses. Few know that in the days when Pacific Grove was a religious encampment, this place was known as "Lovers of Jesus Point." Probably few who gather here now could claim that identity.

Cliffs west of Lover's Point are planted with aloes, yuccas, and another exotic, a close-matting succulent loosely called iceplant, which thrives in this salt-laden air. In spring, it bursts forth in millions of tiny lavender blooms — a spectacle that rivals the landward slopes where poppies and lupines carpet whole hillsides.

Houses small and large line the bay in neat rows all the way to the Point of Pines at the tip of the peninsula, where stands of Monterey pines make their first seaside appearance on the low dunes. Near the sea they are thin and wind pruned, further inland, thick and stout.

The western sea front of the Monterey Peninsula is a place of dunes and golf courses and sea bluffs. The homes along the 17 Mile Drive sit among great pines, gray with the lichen hanging from their limbs. Along the water, pines give way to Monterey cypress, that other tree endemic to the peninsula and its environs. They cling to exposed, granite precipices, where constant winds have twisted them into gnarled forms that look as if some bonsai master had shaped their growth.

They grow old and they falter, their branches no longer able to produce the needles necessary to sustain photosynthesis. Their limbs are stripped of bark by wind and spray, but they stand for many years more as great gray ghosts, dominating the bluffs they cling to. In time, some hard wind driving off the sea brings them low; the downfall of a well-known relict, duly noted in the pages of the *Monterey Peninsula Herald*, is cause for great distress among those who have grown to cherish these giants. Even in death, their scattered white bones cling to the land, disintegrating slowly. At length they return to the earth, where they decompose into nutrient for new saplings struggling for a foothold in the crannies of these inhospitable cliffs.

The southernmost cliff is Pescadero Point, northern terminus of the Bay of Carmel. A century ago Chinese fishermen lived here, but they disappeared long ago. Now ancient cypress skeletons hold sole possession, their bare, twisted branches held high in blessing over Stillwater Cove and the mile-long stretch of gleaming white sand that is Carmel Beach.

In winter, the beach narrows as buffeting storms move its sand out to sea; great rock reefs are uncovered and rise up out of

Sandpipers probe the wet sands for food while
cormorants, fresh from fishing in the sea,
dry themselves on pinnacles high above the water.

the sand that remains. After the vernal equinox, storms recede, and the waves slowly, imperceptibly return what they stole in winter. Sand again covers the rock, and high-tide marks no longer touch the base of the bluffs.

Offshore, huge beds of kelp float on the surface of the bay. Waves tear them loose from their anchorage on the bottom and pile them in great heaps on the beach. Sometimes still living octopi, sea stars, and kelp crabs are tangled in them. Beach hoppers and sand fleas dwell in them, too, by the millions. In springtime, migrating yellow-rumped warblers feast on the small flies that inhabit this detritus of the sea; even black phoebes, small flycatchers that frequent creeks and river banks, leave their fresh-water habitat to partake of the bounty.

This is a place of shorebirds: of sanderlings advancing and retreating with each wavelet on the beach, a *corps de ballet* drilled to perform in perfect unison upon a wet and glistening stage; of drab gray willets that suddenly flash splendid black-and-white wingmarks as they spring into startled flight; of strollers and panting runners, of prodders and pokers investigating every clump of kelp cast up by the sea; of soaked children shouting and wet dogs leaping through the surf in pursuit of birds and thrown sticks.

Those who live nearby and those who come from far places to walk upon this strand are prone to believe there may be no finer beach in all of California.

On dunes facing the sea, Monterey pines grow close and tall; those nearest the coast are pruned and bent by constant winds that blow toward the land.

POINT LOBOS

Point Lobos forms the southern tip of the crescent-shaped indentation in the coast called Carmel Bay. At one time it was a whaling station littered with great bones; at another, a loading point for ships that came to take on coal from mines in nearby Malpaso Canyon; at still another, a prospective townsite laid out in neat grids. There were even days when it was a secluded haven where furtive vessels bearing forbidden elixirs could unload their cargo into caves and inlets hidden from the eyes of government agents. During its last days in private hands, it was part of a dairy ranch, whose owners later conveyed it to the state to become an ecological reserve within the state park system.

Where the road leads from the highway into the park is a place of pines growing in close-ordered ranks. A path leads from the ranger station near the entrance through pines and thick clumps of monkey flower bushes, then circles the ocean front along the bluffs of Whaler's Cove and Coal Chute Point.

In the quiet waters below, sea lions — sportive males tumbling and fighting, females guarding their young — plunge and splash in the gentle swells. Sea otters float in the kelp here, protected from most of those that would harm them.

On the bare headlands facing Carmel Bay, wild flowers in great numbers brighten the boulder-strewn ground — zygadene lilies, seaside daisies, Indian paintbrush, fairy lanterns, flowers of every conceivable color in prodigal abundance. Almost hidden by the overburden of blossoms are *metates*, hollows ground into the living rock by long-departed aborigines who converted acorns into a bitter flour by pulverizing them in these primitive mortars.

Late one sunny afternoon, a bobcat trotting along this trail came to a halt not thirty feet away and faced me for a short eternity. From his mouth dangled the limp, still bleeding body of a ground squirrel. His yellow eyes watched me as we both stood frozen in mid-stride. The dark, vertical slits in his eyes seemed to narrow perceptibly as he waited and considered his next move. The sun, low behind him, outlined each hair on his tufted ears, cocked tensely toward this strange interloper who stood athwart his path.

At this moment, the body of the squirrel

Gray fogs creep in from the Pacific and turn the forest into a place of mists and wraiths.

shivered in some delayed paroxysm of dying and reminded the bobcat of the primary business at hand. He tossed his prey up, caught a fresh grip on it, took one last look at this unexpected obstacle he had encountered, and bounded off the path to disappear into the welter of poison oak and small bushes that lined the trail.

North of Whaler's Cove, past the old, whitewashed remains of some forgotten whaler's cabin, the path climbs steeply up into granite ridges. Pines here make room for enormous, twisted cypress rooted in the sheer face of cliffs that rise vertically out of the sea. The wind has pruned and shaped them in strange ways, and the older, more exposed among them are bare of leaves except at their tips. They look as if they were holding tenuously to the last shreds of life, but this is the nature of these trees when exposed to such harsh conditions. They are tough to a degree beyond human experience. Some of them have for centuries taken all that the elements could hurl at them and still stand hardy and defiant upon these cliffs.

The path goes on through the thick grove of pines above Bluefish Cove, where the trees grow tall and spindly as they compete for light on this north slope. Here the sounds of waves breaking on the beach below are strangely muted, and there is only the gentle sound of the wind moving through the tall pines above. The ground is thick and spongy with the duff of fallen pine needles of uncounted centuries.

Frequently the path makes little detours to viewpoints overlooking the sea. The cypress here, sheltered from the full face of

the sun, are covered with thick red algae that have turned their trunks from white to rust. On the rocky cliffs great clusters of dudleya, a rosette-like succulent, send their roots deep into fractures in the stone to seek out food and foothold.

Offshore lie great, craggy rocks surrounded by boiling surf; here cormorants and gulls and guillemots nest, undisturbed by the prying eyes — both curious and predatory — that watch them across a chasm of restless water.

The trail winds on through the forest until it reaches a vantage point above the huge rocks that rise out of the sea in jagged ridges to form the tip of Point Lobos. Constant winds assault the land here, and no trees grow on these farthermost rocks, nor any plants except those few that take refuge in crannies away from the wind and spray. Lichen mats thick on the boulders, and small ground squirrels come out of tunnels beneath them to beg from travelers along the path.

The way turns back through cypress and pine hung with Spanish moss, descending from the high promontories to meadows along the south shore. The rocks in this area have been tilted up on edge in mixed layers of smooth sandstone and rough conglomerate — rock filled with rounded cobbles that were rolled and tumbled millions of years ago by some river that once flowed here. Where the sea has worn away the matrix rock, these interior stones have been dislodged, leaving the cliffs honeycombed with small, round recesses.

Along these shores, the tide comes in twice a day; cobbles by the hundreds of

On cliffs of Point Lobos, pines and cypress march in weathered ranks down to land's last foothold before the domain of the sea.

thousands tumble in receding surf, and the air is noisy with the sound of their clashing together. Waves pound ledges exposed to the sea, the spray from the impact shooting almost a hundred feet into the air, then falling back to spatter like rain in a winter storm. Sometimes, when a wave comes in undeflected to strike a cliff squarely, water and stone meet with the explosive sound of a heavy artillery piece being fired.

Predictably, the tide comes in and it goes out, but the reach of this advance and recession varies. Sometimes the difference between the highest and lowest tides is considerable, sometimes hardly worth noting. Extremes of more than ten feet have been recorded, but the usual range is little more than half that.

The strip of shore between the highest tide mark and the lowest harbors a swarm of life; probably nowhere else is there such a plentitude of living things large enough to be seen with the unaided eye. This is not surprising when we remember that in all probability every organism on earth can trace its origins back to some remote ancestor who crept from this biome to a drier habitat beyond the reach of the sea.

At the highest level of the intertidal strip — the splash zone, where waves spend themselves upon the shore and then rush back into the sea — the receding tide uncovers acorn barnacles and limpets attached firmly to exposed rock. Around them scurry shore crabs and periwinkles — small, helical snails that are able to move about easily either in water or out. There are also rock lice that resemble their cousin, the common sow bug of California gardens;

they move quickly and keep just beyond the reach of the surf.

Below them, as the tide exposes more and more of the beach slope, are turban snails that never seem to stop moving. Most of the other organisms here are attached solidly to rocks and must stand and face whatever hardship this exposure to air, heat, and dessication will bring. There are mussels and gooseneck barnacles in close clusters; sea palms, short members of the kelp tribe that whip back and forth like trees during a hurricane when waves sweep over them; starfish and bat stars, many-petaled stars of the sea that move imperceptibly on feet lined with suction cups of enormous efficiency; chitons, segmented creatures that migrate across the rock faces but look and feel like rock themselves when the sea withdraws its mantle of protection.

Lovely coraline algae grow here, white or pink plants resembling little branches of coral that the sea has broken apart and scattered everywhere in tiny shreds of great delicacy. Sea lettuce stands in neat green rows upon the flat rocks, and in the niches spiny purple sea urchins lurk.

Even when the tide reaches its nadir, pools are left whose inhabitants are seldom exposed to air — indeed, some probably could not long survive in that alien environment. There are abalone and anemones, octopi and sculpins, hermit crabs and nudibranchs; they hide under ledges, in cracks, or under clumps of feather boa kelp, whose long fronds float on the pools and conceal the life that hides below.

Algae, red-edged with a delicate tracery of white, cluster along small rock faces

Above the pounding surf a shorebird pauses to survey his world — the jagged rocks of Lobos.

below the water line and wave slowly back and forth in the ebb and surge of the tide. Anemones resemble purple flowers but are in reality animals fixed firmly to the rocks, sometimes alone but more often in large colonies; they spread themselves open, tentacles upreaching, and move gently with the swells, alert to snare whatever morsel ventures within the trap of these waving fingers. When the tide leaves the higher ones exposed, they draw themselves inward to become little nondescript olive-green balls, covered with a mucous that catches broken fragments of dried coraline algae. They are camouflaged most expertly until an unwary foot presses down upon them — then they squish and squirt and collapse momentarily under the unaccustomed weight, to regain their shape when the danger has passed.

Sea urchins, with their dark purple spines, scrape out recesses in the rock just below the low-water line; here they live out their adult lives. When they die, their spines fall away, and waves wash ashore their empty tests — hollow skeletons that resemble tiny pincushions with symmetrical rows of pinheads radiating from their centers.

The enemies that prey upon creatures of the intertidal zone are often alien to this milieu; they are not of it but come to feast upon it. At the point of lowest tide, on days when the Pacific is at rest and its gentle peaks and valleys are only a reminder of the enormous force temporarily abated, all the winged foragers of the shore flock here to inspect each exposed ledge to see what juicy delicacy the sea has served them.

There are willets and turnstones, surfbirds and tattlers and jet black oyster-catchers, which are difficult to see against the dark rocks — only their blood-red bills, pink legs, and shrill conversing betray their whereabouts.

In the small sheltered cove of Weston Beach, riding the swells in company with grebes both large and small, are migrating loons on winter vacation from the Arctic. When they dive, they jump almost clear of the water and execute a neat, vertical jack-knife straight down; when they come up, they surface like a jack-in-the-box, always in some unexpected place. These are among the most primitive of birds; it takes no great imagination to note the resemblance to their ancient reptilian ancestors.

Just beyond the ledges that protect this place, great brown pelicans flap slowly and majestically along, their capacious beaks outstretched. Sometimes they sink gradually down to almost touch the sea's surface, then sail along dipping and rising with the swell, their sharp eyes alert for telltale signs of an unwary fish that may venture close to the surface. When they find one, they fold their wings and plunge without grace into the sea. In a moment they surface again, stretch their long bills skyward in a spastic gasp, and the hapless prey slides from the pelican's commodious pouch down his long gullet into his insatiable maw.

Black-crowned night herons fish at the base of the rocks. Great blue herons often perch on driftwood caught in the kelp beds and ride up and down on the roller coaster of the swells, their long necks periscopes keeping sharp watch on the surrounding sea. Gulls are everywhere — California and herring gulls, western and Heerman's, all graceful gliders whose wide wings search out every updraft to ride the air currents that course along this coast.

Seals and sea lions gave this point of land its name. Spaniards who first explored the bay saw them in great numbers, frolicking in the sea, barking and conversing with each other on the rocks, and called them "sea wolves." Primitive Spanish maps noted this headland as *La Punta de los Lobos Marinos,* and so it has been ever since.

The beasts are still here today. Some are sleek and black, some dingy and brown, some gray and spotted. They haul out upon the rocks to sun themselves, and the receding tide often leaves them balanced like a tabletop upon some small peak of stone. There they doze, so attuned to the rhythms of the sea that when waves of more than normal height approach, they raise their heads and tails without looking to avoid the surge.

With great, soft eyes, they watch for enemies, but they have few — killer whales and sharks, mostly. Men no longer hunt them here for food or fur; only an occasional fisherman takes a gun to one of them when it rips through his nets to get at the horde of fish gathered there.

At the south end of the park, the trail climbs high above the sea again, circling two emerald coves with sparking white beaches, and reaches its end on a bare cliff. Beyond it, great bare rocks stand out of the sea, innocent of grass, flower, or tree. Indeed, nothing can grow here because of the thick, white guano deposits left over the centuries by unnumbered generations of cormorants, whose descendants still use it as their rookery, coming and going ceaselessly.

Toward sunset the homing birds arrive in black skeins hundreds of feet long, skimming rapidly only inches above the waves. There are instinctive rules that each bird follows, so that the rock is like a great airfield where the constant arrivals and departures are directed by some unseen controller.

These shores, forests, and cliffs have stood here for millennia — as beautiful a land as lies on America's western shore. Time has wrought no harm; its passage has left only a gentle patina. The place has even survived with grace all the indignities man has visited upon it in nearly four centuries. Trees and brush have grown to cover all the old scars. Now there are only those that are being carved into it today by the tread of too many feet.

Starfish, large and small, inch their way about the tidepools searching for food ...

... while sea anemones, anchored fast to the rock, spread their tentacles and await whatever the sea will bring them.

Sea palms thrash about in the boiling surf like miniature tropical trees in a hurricane.

Day after day, eon after eon, the sea crashes against the ledges of Point Lobos, slowly wearing them away.

Kelp fronds, floating in deep pools between the offshore rocks, swing to and fro with the rhythm of the waves.

Sea otters find shelter in waters surging against coastal rocks.

On islands close to shore are colonies of sea birds that have existed there in safety for centuries.

THE SUR COAST

Properly considered, the Big Sur is a small area — a strip of coast that extends from Point Sur south a few miles to Post Summit, and the valley immediately inland that forms the drainage basin of the Big Sur River. But over the years, the name has been attached to a much larger area; mention Big Sur and most people who live beyond its edges think of the whole coast south of Malpaso Creek at the border of the Carmel Highlands. Those who live in the Sur Valley use the term much more exclusively. But from any point of view, the Sur Coast, extending almost to Hearst's San Simeon far to the south, is a panorama of unique grandeur.

Along most of the coast, waves break against the foot of precipices hundreds of feet high. There are small beaches so inaccessible as to be approached only from the sea. Rock slides are a constant threat; the road hanging upon the sides of the cliffs is often cut by an avalanche of boulders that severs traffic with the world outside until the highway crews with their lumbering behemoths can move tons of debris to clear the way again. Often the repair is more dangerous than the slide itself; more than

one man has been swept to his death when the machine he was driving was carried over the brink by collapsing earth.

On the leeward side of the ocean-facing ridges grow thick stands of trees, sheltered from the cold, drying winds that flow in from the sea. On frontal slopes overlooking the ocean, the hills more often than not appear almost bare, covered with low brush and grasses, but on their tops are stands of tall, dark green trees — giant ponderosa and Coulter pines, California bay, and canyon oak. In summer, when grasses have dried to white gold, the contrast between the trees and sere slopes is spectacular.

In cool canyons, cut by streams plunging down steeply to meet their end in the sea, tanbark oak and great redwoods grow side by side — two trees that seem to seek each other's company. The coast redwood, tallest among the world's trees, reaches its southern limits in these mountains in a canyon near the Monterey County line.

During the closing years of the century past, man made forays into these canyons to lumber the redwood, as he did — and still does — wherever this tree grows from here to southern Oregon. Only a little lumbering

The sun's last rays cast long blue shadows on the beach and touch the swirling surf with gold; many small, hidden beaches are made of sand as black as coal.

goes on along the Sur Coast now, but the profit this great tree brings in the marketplace keeps it in continuous jeopardy. Its companion oak used to be stripped of its bark for tanning hides; boats put in at landings all along this coast to take on bales of tanbark until the very existence of the tree was threatened. But other tanning agents were found, and the oaks were left in peace; now the tanbark oak grows as luxuriant as before industrial man came this way.

Small, narrow beaches lie at the base of cliffs here — beaches of fine black sand ground small by waves tumbling dark rocks fallen from above. White surf is flung upon dark sand and ebbs, creating starkly beautiful tracery upon the beaches. Sometimes whole colonies of sea lions haul out in inaccessible coves, their fur a warm brown against the black background. They lie in the sun just out of reach of the surf, arguing, fighting, sleeping, mating in a haven safe from all intruders except the rare shark or killer whale that sometimes comes among them as they play offshore beyond the surf.

Other mammals only migrate along this coast, setting their course from headland to cape to point, taking the most direct route from Here to There as surely as if some sextant-wielding navigator were in charge. The California gray whale is one. Each December people gather on the headlands to watch the southward passage of the great gray hulks, slowly making their way from the Bering Sea to havens along the coast of central Baja California. There the cows bring forth young in Scammon's Lagoon and other salty shallows farther south.

Once the grays were hunted to the brink of extinction. There were whaling stations at regular intervals all the way from the Alaskan Gulf to San Diego, and whaling ships brought in whale carcasses by the thousands to be flensed to make margarine, dog food, whale oil for lubricating delicate machines, and a score of other products.

Now they are protected by Mexicans, Americans, and Canadians alike, and to some degree by international agreements with other nations that still harvest whales. Scammon's Lagoon is a Mexican national park where whale watchers are kept in check so that the animals are not disturbed during their mating and birthing activities. And in American waters whalers no longer harpoon the beasts and tow them ashore to be rendered at Point Lobos, Moss Landing, San Francisco, and all the other ports that used to traffic in this commerce.

In March and April they go north again, this time with new calves shepherded by their mothers. They do not hurry; there is time aplenty to make deep soundings, to burst from the sea and fall back with mighty splashes, to blow their breath in white geysers on the cold ocean air. Sometimes they approach the shore so closely to investigate coves and inlets that people living nearby can hear the sounds of their presence.

Perhaps their numbers will increase so that they do not suffer the fate of their cousins who once roamed this coast in great frequency — the great blue whale, the humpback and the finback and the sperm, all of whom seem doomed to disappear from the earth at the hands of Russian, Japanese, and Norwegian whalers.

One day, from their home on rocks above

After the winter rains have come, stream mouths break through sandbars that during dry months blocked their way to the sea.

At the base of high cliffs along the Sur Coast, where the water lies turquoise in shallow, sandy coves, sea froth plays in delicate fronds of kelp.

Early morning light shows Big Sur beaches in another mood; for years man has searched for pieces of jade cast up on these sands by waves rolling in from the open Pacific.

the surf, where a loving vigil has long been kept for living things of the sea, a couple saw a great seal making his way through the waves. He was of more than normal proportions, and as he lumbered closer through the surf, his great protuberant nose revealed him for what he was — that Cyrano of the sea, the elephant seal, not seen in these waters for years. His kind had been reduced to a small herd hiding on the beaches of Guadelupe Island off the coast of central Baja California. Now no longer hunted for their pelts, they have increased in numbers and pass this way each year, bound for or returning from their breeding grounds on Año Nuevo Island, fifty miles to the north.

Yet all these wild things fade into insignificance beside the impact that one small furred animal has made on the history of the American West — and, for that matter, on the history of the entire western world. For only the finding of gold flecks in the race of Sutter's Mill even approached the discovery of the sea otter as a determinant of history along the Pacific Coast.

Before European man established along this shore his first permanent settlements north of Mexico, Russian explorers under Vitus Bering had penetrated the Gulf of Alaska, the Aleutian chain, and eastward along the mainland coast. They saw a multitude of fur-bearing animals — seals, sea lions, and sea otters — and duly noted what they had found.

Their expedition came on bad times, however; scurvy thinned their ranks and felled Bering himself before they could beat a way home through winter storms. But a few survivors reached Siberia and carried news of the rich new land to St. Petersburg. Other expeditions of exploration and exploitation quickly followed, as always happens when word spreads of ripe fruit awaiting the picking.

Russian sea captains took otter pelts back to the court of Empress Elizabeth. The response was instantaneous and profound, for here was a finer, thicker pelt than any ever seen by Europeans — a trade item likely to make its procurers wealthy. The rush was on. Russian crews harvested the otter whenever and wherever it could be found — from the frigid waters of the Aleutians to warmer seas off Alta California.

News of such a treasure spread fast and far. Enterprising British, French, and American captains soon followed the Russians into these waters to pursue the abundant otter. The Russians were the most zealous in putting down roots in the unclaimed lands from Alaska down the coast to northern California, where the Spanish Crown had proclaimed dominion two centuries before. They established settlements, outposts of Tsarist empire, as far south as the Sonoma coast, where Fort Ross still exists with its timbered stockade and an onion-domed church topped by the double cross of the Russian Orthodox faith.

In due time, intimations of these and other incursions into Spanish domain were carried back to San Blas, Loreto, and Acapulco by ships of all flags that plied these waters. Eventually the word reached Mexico City and distant Madrid. It was because of this that the Viceroy of New Spain began to set in motion plans that sent

Sandy beaches are not frequent from Point Lobos south; most of the coast is rock and boiling surf.

In summer, huge banks of fog move in from the sea to draw a blanket of mist over the coastal mountains.

Portolá and de Anza and Serra northward to establish footholds of Spanish dominion from San Diego to San Francisco Bay.

Had the rich harvest of otter skins not precipitated this Spanish investiture, the Spanish place names that are a hallmark of present-day California might have been Slavic instead, and the Russians in possession of a landhold not so cheaply sold as Alaska.

Even so, Russians and others continued to harvest the sea otter. The Americans, when they preempted the land and asserted their sovereignty over the treasures that were here, saw what others had done and continued with avidity to pursue the same or even a more accelerated course. Before the present century was one-quarter past, the southern sea otter was added to the list of other species forever departed: the great auk, the dodo, the passenger pigeon, the Carolina parakeet.

Then, during World War II, a submarine watch on a cliff some fifteen miles below Carmel, where Bixby Creek enters the sea, noticed a movement most strange in the waters below. Report was made to Monterey, and savants from Hopkins Marine Laboratory went down to see what curious thing disported in these waters to distract the sentinels posted there to sound an alarm if the Japanese should dare to approach too close. The professors looked and doubted, argued among themselves, looked again, and finally agreed that indeed sea otters were there, and in force.

For a quarter of a century, this remnant herd had hidden itself along the inaccessible south coast and managed to survive in peace, diving for sea urchins or other delectable morsels from the sea. Where the booty of its plunge was some hard-shelled mollusk, it used a tool to open it — a stone from the bottom, held on its chest, on which it pounded the shell until it opened. The sound of this cracking maneuver, echoing like small pistol shots above the sound of the waves, was probably what led to the otter's rediscovery.

Now protected by stern laws that forbid their molestation, otters have slowly increased their numbers and their range. Most of them are gathered offshore between Monterey and the Point Sur light, but their advance patrols reach out as far as Pismo Beach to the south and Año Nuevo light to the north. Commercial abalone hunters detest them and blame them for depleting the abalone along the coast; in truth, the otter does dine on abalone, but it much prefers the plentiful sea urchin. The otter, by his eating habits, in fact, performs a beneficial service — his staple food, the sea urchin, is a destroyer of the native kelp. As the urchin is harvested by the otter, kelp flourishes and provides a necessary shelter for game fish that, lacking such a haven, would not be here. The otter probably does less damage to abalone colonies than those hunters with scuba diving gear who scour the rocks and pry loose every abalone they find whether of legal size of not. For thousands of years, the sea otter has coexisted with abalone without depleting it; man's record is much more tarnished.

The otter does face other enemies. Sharks and killer whales slash at it, the shark for food, the killer whale for sport.

Ancient marine terraces, lifted up long ago out of the sea, are home now to grasses and flowers — and an occasional cow.

Occasionally a skin diver spears one under water for reasons known only to himself. But its worst enemies are officials of state agencies who listen to loud complaints from abalone gatherers and Pismo clam diggers and lay plans to trap the otters and transport them elsewhere.

Fortunately, the otter is not friendless. He has a powerful defender in a great lady who lives high above the sea in Big Sur; she has connections in important places, friends and followers whose voices count in the halls of government.

Most people who live near the sea here and the thousands who come to see the otters in their native habitat know little of such behind-the-scenes activity, either on behalf of or against the animals. Most are content simply to know they are there, rolling and tumbling in the surf or lolling in offshore kelp beds with their young — a na-tional treasure rediscovered.

Waves cut away inexorably at the under-pinnings of the Sur Coast cliffs and earthquakes are not strangers here. Yet as these forces of nature work to wear away the great ramparts that border the ocean, the movement of the Pacific plate a few miles to the east is raising them farther out of the sea. Natural forces do less to alter the coast than does that late arrival — man. But in spite of his despoliations and dep-redations, the south coast for the most part remains majestic and unspoiled.

It has been said that the drive along the cliffs of this coast is equal to the renowned Amalfi on the Bay of Naples. But those who live on the Sur Coast's ridges and in its valleys and upon its precipices — and many who have passed through, paused and then tarried longer — are sure that it has no equal in all the world.

The lighthouse at Point Sur stands lonely on a rock rising out of the sea beyond the wind-built dunes.

Dunes along the Little Sur River Beach are slowly disappearing under a steady invasion of succulents.

Old sycamores, old roads, old fences are common along the last miles of the Big Sur River before it empties into the sea.

Pampas grass, an alien from Argentina, grows thickly on slopes where soil has been disturbed; here half a mountain slid into the sea during an earthquake midway through this century.

THE MOUNTAINS

South of the Carmel River valley looms the granite and limestone bulk of the Santa Lucia Range, rising from sea level to heights beyond six thousand feet. On its eastern flanks high ridges subside into rolling hills that border the western edge of the Salinas River valley; on the west, steep slopes plunge precipitously into the Pacific.

The range stretches south more than one hundred fifty miles until its ridges are lost in a tangle with the eastward-trending La Panza Range near San Luis Obispo. At its north end, it slips under the Sierra de Salinas, a short transverse range that is of a different period. Some men who know of such things believe that the Farallon Islands, which lie fifty miles off the Golden Gate, are the last peaks of the Santa Lucias before the range disappears forever in the deep trenches of the Pacific.

When this land was raised out of the shallow seas that covered central California during comparatively recent geologic time, upward thrusting of the earth's mantle created furrows that paralleled the great Sierra Nevada a hundred miles to the east.

For the last ten thousand millennia or so, water and wind have scratched and clawed at this island mass, the detritus-laden streams cutting canyons through the long ridges as they coursed to the sea. Their stream gradients were so steep that most of them roared directly into the ocean instead of building flood plains near their mouths. Even today there are silvery waterfalls that leap gracefully from high ledges into waves beating against the cliffs below.

In a few places silt was deposited in large quantities, and thick alluvial fans were built up. As the two great continental plates that oppose each other two score miles to the east have slowly lifted the earth, these fans rose out of the sea and became wide marine terraces, today meadows that slope gently toward the ocean.

As the land eroded away, lichen took hold on bare granitic crags atop the ridges and on limestone reefs raised from the sea. Dying, they left a sparse humus in cracks and crannies, where seeds blown by the wind or left by far-ranging birds took root and began to change the face of the landscape. Trees appeared, cone-bearing at high elevations, deciduous in canyons and on lower slopes. Grasses and small flower-bearing plants made their appearance, covering the

In the fertile soil of the valleys, wildflowers
— crown jewels of the California landscape —
adorn the earth.

hills with a bright mantle of color. To this day, after the soaking rains of winter, wildflowers bloom in a profusion matched only by the desert in years of unusual wetness.

Soon new species began to develop through evolution and mutation within the complexities of this varied biota. Experts who search out and catalog such things have discovered plants and shrubs and trees endemic to these particular mountains and valleys — they grow of their own volition nowhere else in the world.

As the land prospered and more seed-bearing plants appeared, there came also small creatures whose existence depended upon the seeds and roots: mice, voles, shrews, gophers. They were followed by larger creatures, predators: bears and badgers, mountain lions and bobcats, wolves, coyotes, hawks, and snakes. Most have roamed these hills ever since. People who live in the mountains seldom encounter lions, but they occasionally hear them scream in the night — unmistakable and chilling the first time heard.

In recent times even wild boars, whose ancestors were released on ranches above the mouth of Carmel River, have spread this far. Some have mated with domestic pigs to produce a formidable feral porker. One man of many years, the descendant of one of the earliest settlers, keeps his eye on a wild sow that forages on his land above Lucia, feeds her, and wards off her enemies. Now old, she looks upon this land as her private domain, and stands her ground truculently, surrounded by her small progeny, her eyes glinting in suspicion when interlopers approach. One walks warily in large circles around her to avoid confrontation.

These mountains, rising as they do out of a restless sea, act as barriers to approaching storms born in the great disturbances around the Gulf of Alaska. Such storms are part of the huge cyclonic whorls that circle far out into the northern reaches of the Pacific before swinging back to approach the coast from the southwest. When they encounter the mountains, their moisture-bearing clouds are forced to rise and lose their burden. Slopes of the Santa Lucias facing the ocean have recorded annual rainfalls surpassing fifty inches.

Deep in the porous dolomitic heart of the mountains, water is stored by the billions of gallons. When the rains weaken and depart, springs burst forth in great profusion, and creeks are full to overflowing. Even when summer comes, when temperatures rise and forests and *potreros*, those golden mountain meadows, are dusty and tinder-dry, there are still seeps and springs that give life to small creatures that must have water to exist.

After the land was first formed by tortured movements of the earth, three elements — water falling from the sky, wind blowing from the ocean, and fire sweeping over the slopes — were the prime factors in shaping the mountains and controlling the environment.

Lightning is not a thing of regularity during coastal storms; indeed, the spectacle of lightning flashes and the sound of thunder is infrequent enough to be noted with considerable exitement by those who live here.

In winter, sycamores lining streambanks in the Santa Lucias drop their leaves to stand bare and exposed until spring stirs them back into life.

In autumn, however, when the land is at its dryest and the equinoctial storms of September arrive with fireworks from the sky, bolts of lightning loosed upon the earth have more than once started fires of major import. The last in recent years was of monumental proportions: large portions of the forest were blackened by a holocaust that began when lightning struck a stand of ponderosa on Marble Peak overlooking the sea and ranged unchecked for days until it petered out on firelines established far distant on the last ridge above Carmel Valley.

Before modern man arrived on the scene, fire was commonplace but on a less catastrophic scale than now. Underbrush was cleared out often by small, fast-moving fires that roared through at ground level; older

When forest fires rage unchecked in the Santa Lucias, the sky above the peninsula many miles away grows murky with their smoke, and the sun shines hardly brighter than the moon. When the flames die out, the slopes are a desolation of charred branches and scorched roots; only time can erase the scars.

Deep in the redwood groves, ferns and sorrel grow lush, wake-robins thrust up through litter on the forest floor, and toadstools proliferate on damp, rotting stumps of trees.

and larger trees were only slightly damaged, for flames did not linger. But now, aerial tankers drop tons of bright red borates, and teams of firefighters throw up barriers that starve the flames and contain them in their infancy. As a consequence, underbrush grows thick, branches and whole trees fall, the ravines and gullies are choked with debris. So when the fires do come, they find enormous quantities of waiting fuel. In the latest fire, temperatures in some of the canyons at the height of the firestorm exceeded a thousand degrees. Whole forests exploded into fireballs and nothing was left standing. Rare bristlecone firs were destroyed. Animals that could neither flee nor burrow beneath the burning ground-cover were scorched to death; others suffocated when all available oxygen was consumed by the flames. It will be years before the land recovers enough to erase the scars left by this one great fire.

Yet, the most damaging impact to creatures of the wild has come not from wind nor water nor fire. Western man moved into these mountains only a century and a half ago and brought with him beasts that preempted the habitat of native animals. To protect his cattle and sheep and goats, he hunted the wolves and grizzlies to extinction. The mountain lion has survived only because of his overwhelming shyness — he kept a low profile — and because determined people worked to remove the bounty that was formerly paid to those who reduced a lithe, bounding cat of enormous grace to a bloody, mangled trophy.

Once in a while a straying calf does fall to a hungry cat, but mostly the lions feed on deer — the old and the sick — and on rabbits and other small rodents.

The coyote, too has held on, for the same reason he has survived and increased everywhere in spite of the most determined efforts to effect his extermination — he is smarter than those who would do him in.

Man is slow to realize that these mountains are the better because of wild things — pumas and bobcats, deer and coyote, badger and eagles, and all the other creatures. Their passing would diminish the land forever.

In summer months, cool fogs hang off the coast until warmer air over the land rises and draws them in from the sea.

THE VALLEY

Far up on the mile-high slopes of Chews Ridge in the narrow upper reaches of Miller Canyon, a cluster of small springs marks the headwaters of the Carmel River. In wet years the springs flow copiously; in years of drought they shrink to the edge of disappearance. Some may dry up completely, their underground stores exhausted. But always one or two keep flowing, so the upper river never goes dry through natural causes.

Like most rivers in the American West, the Carmel doesn't start out as much — a steady drip under a mossy rock, a small trickle that starts on its way down toward the sea to be joined by other trickles from other springs lower down. Tributaries join it, increasing its volume to a brook, then to a stream too wide to jump, knee deep, noisy with arpeggios of water foaming over and around white granite boulders in its path.

Green reeds grow thickly along its borders, ferns native to these mountains overhang it, and sycamores that seek its moisture spread huge roots along its banks. Small black ouzels flit from twig to rock to sparkling shadows — skindivers of the thrush family in search of helgramites and other delicacies that live here in the bottom of the stream; their feathers, wet from continued forays into the water, glisten in the sunlight.

The river, in these upper reaches, runs swiftly through narrow valleys formed by precipitous slopes of the Santa Lucia's higher ranges and sets a direct course for the sea. Then the mountain bastion dominated by the granite slopes of Ventana Cone and Ventana Double Cone bars the way and turns the river aside in a long horseshoe heading north away from the sea. At length the flow spreads out into muddy waters caught behind Los Padres Dam, first of the reservoirs that serve the people and towns downstream.

Just below the dam is the confluence with Cachagua Creek in a lovely valley a mile wide and a dozen long, parklike with its graceful oaks, sloping meadows, and pastures with grazing Herefords and horses. The river meanders through the Cachagua valley, but it no longer runs free. The dam stems its flow until storage reservoirs downstream fall too low; then the water flows again but only because distant Carmel, Monterey, and other communities

Streams that disappear in dry months run noisily in winter, when the land is replenished by rainstorms from the Pacific.

Great gnarled oaks survive for centuries; when they fall, their skeletons still persist for years.

look to this slender stream to fill their needs. Only during the winter rainy season does water run freely over the dams and to the sea. After the storms move north in the late spring, the great Pacific high pressure area settles in to parch the land and bring fog to the shore. Every drop that comes down from the high springs and tributaries is caught and hoarded by the Los Padres and the second dam downstream, the San Clemente.

Mostly the river proves adequate to the demands upon it. But there are dry cycles, years when rain falls seldom or not at all. Springs disappear, the river ceases to flow, waters behind the dams slowly shrink until the lakes become dry, cracked mudflats filled with the snags of trees killed by deepening waters when the dams were built years ago.

Water becomes a precious thing in mains of the towns; rationed, it is a resource to be hoarded, to be eked out by the quart, and the land becomes a place of gardens withered, cars unwashed, baths untaken, toilets unflushed. Yet, because that most coveted of all prey, the tourist, must be courted at all costs, motels are allowed generous allotments, and perspiring locals complain bitterly about outsiders who luxuriate in their showers while scarce water runs down the drains in prodigal waste.

At those times there are moves afoot to limit the call on this scarce resource by curtailing new construction. Carpenters, electricians, and all the rest of the building trades go fishing or take up other lines of work. Queues at the unemployment office grow long. There is much cursing and com-

plaining, and the situation is all the more frustrating because there is nobody to blame for the drought — except perhaps the water company for not making the dams higher, pipes bigger, or wells in the aquifer deeper.

In time the rains come again — they always do. In normal years, Pacific storms beating against the coast bring as much as forty inches of rain to these slopes during the wet months. There are storms — and not infrequently — whipped off the sea by hurricane winds, and gray clouds, fat with rain, obscure the face of the land. Tons of water fall in the space of a few minutes; the storm relents and then strikes anew, time after time, so that the cascades of rain run off the water-soaked land to swell brooks into raging torrents, creeks into roaring rivers. Trees along the banks of the Carmel are torn from the soil and roll tossing and heaving upon the flood, battering all in their path; they pile up against bridges until the spans collapse under the weight of the water dammed up behind them. Beaches near the river's mouth are littered with trunks and branches that have been scoured clean by the churning of the river. For weeks afterward, people who live nearby hack away at free firewood piled high upon the sands.

The river gnaws at its banks and eats away huge chunks of earth that splash into roiled waters, adding tons of silt to the river's already muddy burden. The streambed changes; valley people who have built beside it stand anxiously and helplessly by while the flood claws at the underpinnings of their homes.

Enduring oaks and ephemeral flowers bedeck the hills that line the upper Carmel Valley.

Once the rainy season has raised the reservoirs to running over, water flows briskly for half the year. Steelhead trout — homecoming from life spent somewhere in the open sea — spawn in the shallows upstream. But by early summer the river is dry and sere. In the last deep pools left upstream, huge trout await their end; the waters will dry up and leave them marooned in muddy shallows, prey to bobcat or coyote or marauding racoon.

Downstream, past Robles del Rio and Rosie's Cracker Barrel, past the River Ranch and the Boronda Adobe, past Garland Park and the Farm Center, where the river cuts around the shoulder of the last transverse ridge and heads straight for the sea, the valley widens out into flat, orderly fields, where truck farmers hold out in a last desperate defense against the tide of building that threatens to turn the land into a vast housing tract.

The river runs on through three golf courses in the lower valley, passes by the sprawl of shopping centers that borders the Odello artichoke fields on the flood plain, and debouches into marshes that mark the lagoon formed when the sea casts high sandbars across the river's mouth.

In autumn migrating ducks and geese flying along the coast from Alaska to warmer winters in the south put in here to feed on the grasses and marsh plants that fill the lagoon. Teal and bufflehead, canvasback and redhead are here in force; some birds stay for the winter but most fly on. There are some mallards that never leave; they mark this territory as their own and stake their claims against all interlopers.

In the time of the mission fathers, the river always ran to the sea; not until the first dam was built by the Pacific Improvement Company to water the grounds at the first Del Monte Hotel did the streambed run dry. Now, through most of the year the river never reaches the lagoon — its waters are all diverted. But when the rains come, the dry bed becomes a roaring sluice; the shrunken lagoon again grows wide and deep until the river breaks through the bar and the blue waters of Carmel Bay turn red with the mud of distant hills.

White star lilies, blue fiesta flowers, creamcups and ubiquitous California poppies splash the green slopes after winter rains quicken the land.

THE CHANGING COMMUNITIES

*On cool evenings, when the sun sinks low,
flames crackle in fireplaces throughout Carmel,
and the smell of wood smoke fills the air.*

Monterey, Pacific Grove, and Carmel came into being for very disparate reasons — Monterey out of geopolitical necessity, Pacific Grove for religious reasons, and Carmel as a real estate sales operation. Some cynics hold that it is still primarily a real estate promotion of inflated proportions, others that it is simply mercantilism run wild.

All three towns have existed for eminently respectable lengths of time by American standards — Carmel is approaching its centennial, Pacific Grove was a settlement more than a hundred years ago, and Monterey is surpassed in antiquity along the Pacific Coast only by San Diego.

During and after World War II, small satellite communities sprang up on the borders of Monterey and Fort Ord. Through the years they have grown so rapidly that Seaside is now larger than Monterey, and Marina soon may be.

But these are raw, youthful towns, full of growing pains and rough edges. All the new communities — Seaside, Marina, Del Rey Oaks, Sand City — were nothing more than brush-covered sand dunes forty years ago. Like many California towns of the same stripe, they never really grew; they simply exploded. Perhaps in time, when the dust has settled, these towns, too, may develop personality.

The three older towns were separate entities from the start, not intended to serve as bedroom or service communities. Each is largely sufficient unto itself. To this day, each is a little suspicious of the others and highly jealous of its own independent stature. They do not find it easy to work in concert, and each has peculiar views concerning the people of the other two.

Because this is so, the three towns have only slight resemblance to each other. Monterey still retains the air of its Mexican antecedents, and Pacific Grove the turn-of-the-century look of a town built by New Englanders, while Carmel defies description. Some say — as has been said of the whole state of California — that it is less a place than a state of mind. They may be right.

Most of those who live on this peninsula came from some other place. Some came because the military brought them, others because they vacationed here. For whatever reason, they came back to stay, joined by many more who simply migrated West during the great immigration wave of the last thirty years.

Residents compare notes to establish dates of tenure here, and a status of sorts is conferred on — or assumed by — those who boast the earlier dates of arrival. An amazing number who came here as strangers in their more youthful years have remained and, with the passage of time, slipped almost imperceptibly into that typically American role — the "old-timer."

On the Fourth of July, people come from all the neighboring towns to celebrate on the beach in Carmel.

MONTEREY

On June 3, 1770, six years, one month, and a day before John Hancock and his friends put pen to a document that marked the beginnings of the American republic, Gaspar de Portolá for the Spanish Crown and Junípero Serra for the Roman church stood beneath an ancient oak on the southern shore of the great bay they had traveled two thousand miles to find and proclaimed the establishment of the outpost colony of Monterey. No other community on the Pacific shores of America — save for San Diego, which predated it by a few days less than a year — is older.

That this is so is a source of great pride and comfort to many present-day residents of Monterey. Some lay claim to forebears who came here as soldiers in Portolá's command, or as impoverished settlers in Juan Bautista de Anza's entourage, fleeing hard times in Sinaloa. Others have been in Monterey hardly longer than yesterday; but because restless Americans have few permanent roots, they often tend to seek out institutions of relative antiquity with which to establish some tie. Just as converts are always the most vociferous evangels of their adopted faith, the shakers and movers of the Monterey History and Art Association — the group that nurses carefully the claim that Monterey holds on the primal history of California — for the most part came from somewhere else.

But the fact remains that for seventy-nine years, under the dominion of Spain, Mexico, and the United States, Monterey was the capital of California. Then, when the fever of the rush for gold in the Sierra Nevada foothills in 1849 left the town almost deserted, populated by women, old men, and children, the center of activity moved north — business and shipping to the new, almost exploding town of San Francisco, the capital to San Jose.

Monterey never regained its old flavor. Those days, when the candlelit *salas* in the adobes of the Malarins, Soberanes, and Estradas were bright with laughter and soft Spanish phrases and the music of guitar and fiddle, lived on only in the memories of the faded *doñas* of the old families. Even the families themselves were disappearing — pushed aside by American newcomers lured here by unlimited opportunity. *Yanquis* married the daughters of the gentry and became the new owners of the old Spanish

No longer is the harbor at Monterey jammed with large boats of the sardine fleet; the sardines are gone. Now there are only smaller fishing vessels and pleasure craft.

In the marina, immaculate pleasure boats rock in the swells beside battered salmon trollers and tuna boats.

land grants. Those not so lucky in love resorted to more devious means to euchre the guileless *hidalgos*, who were no match for Americans playing with a different set of rules. It was not long before the Mexicans, who had taken the land from the Indians, lost it to the new masters of California.

Nowadays one walks along the clean, swept streets of Monterey and through the grounds of old adobes maintained by the state and imagines that one has stepped back into a place and time that have been gone for more than a century and a quarter.

The romantic past seems to linger in every old rose garden beneath ancient balconies. Perhaps — but photographs and drawings from that day present a picture not quite consonant with well-kept walls and gardens of today. There were certain amenities lacking, disorderly animals in constant residence, and attendant aromas incident to that time. During the rainy season, the streets were a morass of mud. Gardens, when they existed at all, were not nearly as luxuriant as now because water had to be laboriously carried by hand to nurture them — by people not noted for their industry.

During the century that has followed, the beauty of Monterey and the land around it has been a magnet drawing so many people that the number who have remained now threatens to change the place into a replica of crowded metropolitan areas lying farther north.

That it has not already changed beyond recall is probably due to two men who came here in earlier years of this century — S. F. B. Morse and Allen Griffen. Neither of them lived in Monterey — in fact, both lived within the guarded grounds of Pebble Beach — but their influence was felt far beyond their immediate precincts.

Sam Morse, grandnephew of the original S. F. B. Morse, early American painter and inventor of the telegraph, was for half a century the overlord of his Del Monte Properties Company, which owned the choicest of the peninsula's land. He saw early the beauty of these cypress-bordered shores and pine-covered hills, knew their commercial worth, and made sure the land was treated gently. He had a special interest in Monterey — his world-famous Hotel Del Monte and its magnificent park lay within the limits of the town.

A most powerful and persuasive man, he exerted a strong influence on the governments of the county and surrounding cities — mostly by suggestion and by providing help of one kind or another for them to go in the direction he wanted. He was not primarily an altruist — indeed he was probably no altruist at all. It was simply a case of enlightened self-interest that benefited every community that surrounded his dukedom and kept them from unsightly sprawl.

He was aided, sometimes intentionally and sometimes indirectly, by Allen Griffen, who owned the *Monterey Peninsula Herald*, the region's only daily newspaper. Griffen's voice was a powerful one, which commanded respect and attention far beyond his local audience. Both Morse and Griffen were listened to in the golden-domed statehouse in Sacramento and even the halls of Congress.

Migrating birds from the entire length of the Pacific Coast stop to rest in the protected waters of Monterey harbor.

Both believed in growth as a natural and preferable condition of American enterprise, but both knew that uncontrolled growth can be a cancer upon the land. Their actions and influence, more than any others', set the tone of the place: an urban forest where trees took preference over surveyor's transit or developer's bulldozer. Both men are gone now from the public scene — one dead, the other retired to his garden. The marks they left upon the land are the monuments that point to their passage.

Since the time Americans took ownership of the land, the pine slopes of Mon-terey have attracted artists of one stripe or another. The first of real note was Robert Louis Stevenson, who traveled coughing and wracked with illness from Scotland to New York, then on across the country by immigrant train to Monterey, following his love, Fanny Osborne, wife of another. He found her here and waited until she divorced her husband, a most scandalous procedure in that time. He lived on the fringes of starvation until his disapproving family in Edinburgh took pity and sent him a stipend until his health improved and his royalties increased.

He took his meals in the hostelry of Jules

The harvest from the sea comes in each day, caught by men who fish as their Sicilian ancestors have done for a thousand years.

Simoneau, in the company of his French host and certain of the town's leisure gentry. He explored the woods and the shores, walked the cliffs and climbed the hills, and wrote about them all. There are those who believe, with probable reason, that the rugged domain of Point Lobos was fixed firmly in Stevenson's mind when "Treasure Island" was set to paper, long after he had departed the peninsula with his beloved Fanny.

Stevenson was not the first of his breed to stay awhile beside the bay and memorialize it in word or brush stroke — Jules Tavernier was working at his easel here when Stevenson arrived in the 1870s. And he certainly was not the last. By the beginning of the twentieth century, painters from the East were coming out to join Charles Rollo Peters, that painter of moonlight on Monterey adobes — a parade that has never ended although their reasons for coming in later years may be somewhat different.

During the century that began with the raising of the American flag over the Customshouse, the *paisano* largely disappeared. His place was taken by Americans and Sicilian immigrants, who came with scores of relatives to man the sardine fleet that plied these waters. The economy of the town was tied largely to the plentitude or scarcity of the sardine. Fortunes were made by the cannery owners and by astute moneylenders who financed the fishermen, holding their boats and expensive nets as security. The fishermen made a living but none grew rich — not from fishing.

In the old days not all the people of Monterey lived in the small, well-kept houses of the Sicilians nor in mansions on the oak-shaded slopes overlooking the bay. Another stratum of society inhabited the empty spaces between the cannery pilings, discarded crates in vacant lots between the warehouses and the whorehouses, or the rotting and upended boats that were abandoned above the highest reaches of the winter waves.

These were the floaters, those whose way of life seldom varied in good times or bad. The pursuit of sustenance more often than not took a back seat to the incessant search for that most prized among earthly possessions — a bottle of muscatel. Occasionally when life was more generous, they found solace in cheap bourbon, but for the most part red wine of little quality but sufficient quantity was the staff of life to this brotherhood.

Their antecedents were probably here shortly after the first Spanish settlers. Robert Louis Stevenson took note of them when he was in Monterey a century ago. Then, in the Depression years, a tale-spinning refugee from Salinas and Stanford, John Steinbeck, arrived and moved among the shadows of this demimonde. With finely tuned ear he listened to their talk, and with rare perception examined their lives. He made their stories his own; he embellished and expanded them — all with gentle humor and sympathy. The *paisanos* of Tortilla Flat and the motley denizens of Cannery Row stepped out of the pages of his books and took their places as mythic heroes beside Tom Sawyer and Huck Finn, Elmer Gantry and George Babbitt, Jay Gatsby and Ichabod Crane and Hester Prynne.

When fishing boats unload their catch,
gulls mount a careful vigil for dropped morsels
or scraps thrown overboard.

Among this breed was one known simply as Cockeye Romero. Undoubtedly he had once possessed a name given him by his parents at the time of his birth, a name after one of the saints of a Church in which he no longer maintained the odor of sanctity. Even this name, which aptly described the way in which he looked upon the world — that is to say, in more than one direction at a time — this name also slipped into disuse, and he became known from one end of Monterey to the other simply as Pilon. In the language of his ancestors that meant "something for nothing," a *lagniappe* or largesse bestowed by the more fortunate upon the less. Not that Pilon was ever such a bestower; on the contrary, he spent his life in search of such gratuity. That he was eminently successful was proven amply by the fact that no one of his time had ever known him to be employed other than nefariously.

He had a friend, one like himself in habit and demeanor. Whether he was a *paisano* or not is lost in the fog of years. Most likely he came into the world as Juanito Martinez but the police blotters of that time named him simply Johnnie Martin, and so he was known to all of Monterey.

They were inseparable, Johnnie and Pilon; theirs was a friendship marked by deep ties of kindredness. They stole together, drank together, fought together — most often with each other. After a night of serious encounter with the grape, they frequently arose in the morning broken in body both from the effects of massive imbibition and from knife wounds each had inflicted on the other. But such honorable scars were stoically endured; indeed, they

scars were stoically endured; indeed, they were even worn with pride, much as students at Heidelberg gloried in the saber slashes on their cheeks.

They spent time in jail, and such was their attachment that mostly they were there together. When they were not, their enforced separation lay heavily on the one left free and alone, for men standing apart from their fellows, defiant, need aiders and abettors to stiffen their spines and bolster their resolve.

They made a home together, of sorts. In Iris Canyon, not far from where the college now stands, they found hollowed into the face of the hillside a cave that seemed eminently suitable to their needs. It provided a shelter overhead and had no apparent landlord; further, there grew around it enough poison oak to discourage timid souls with more tender hides.

With scraps of wood and cardboard and tin they built a precarious wall behind which they could retire and find a degree of protection, from the elements and from the disapproval of the world outside.

They had no need for amenities of higher degree — water is mainly needed in quantity when one washes, electricity when one reads. There is no record of such endeavors by either Pilon or his friend. And the cave was set far enough apart from the high roads of ordinary commerce that it was not a thorn in the side of the health authorities.

Wood was nearby in plenty to be picked up for fire. What food they ate was harvested here and there without the legal owners being the wiser. Sometimes, when luck was bad, there was nothing to eat, but the pangs of hunger were endured stoically until another day. At other times, however, their fortunes fell below the level of human endurance — there was not muscatel. Without food one merely hungers, but without wine the whole world loses its luster.

The valley below the cave was testimony to their persistence, good fortune, and occasional blind luck in securing this necessary provender. The slope was paved with green glass shards — broken wine bottles hurled from the mouth of the cave to splinter on impact below.

But Pilon and Johnnie, when the fog subsided and the weather turned balmy, often abandoned their customary place to live in the open — probably because in such warm weather the fleas they lived with drove them out. So it was they took up residence one warm September in a gully behind the cemetery — a place with many bushes to provide privacy and protection. They were observed there roasting some purloined fowl, late of an unwatched Monterey henhouse, and sleeping off indulgences. No one would have thought overmuch about such an occurrence had not Johnnie and Pilon taken to filching wooden markers from the cemetery to fuel their culinary fires. These ornate, wooden crosses marked Italian graves — Sicilian graves, at that. So to head off an inevitable vendetta, the police sent an envoy to put the fear of God into the culprits. The agent chosen for this task was one Albert Elasho, a policeman wise in the ways of *paisanos* of the stripe of Pilon and his friend.

Rather than mount a frontal assault upon their place of retreat, Elasho stole upon them through the cemetery and took up his vigil in a thicket close by, almost within arm's length of the pair, by this time well down the path toward alcoholic nirvana. One bottle lay empty by a dying fire, a second bore the stamp of serious inroads.

To replenish their fire against the chill of approaching evening, Pilon ventured into the cemetery once more, returning shortly with two wooden crosses he had uprooted from the earth. He slid down the bank quickly after sweeping a bleary eye around to see if there were observers to his malfeasance. Satisfied, he tossed the wood on the fire.

The jug of wine was duly demanded by Pilon and grudgingly given. Pilon lifted it, judged its color and quantity, raised it high, and tilting it toward his dry and eager lips, began to drink with great, gurgling gusto.

At this precise moment, Elasho, concealed in the bushes hardly more than a yard or two away, drew his Police Special .38 and neatly shot the bottle out of Pilon's hands.

An awe-struck Pilon stood there, his empty hands held high, purple wine cascading down his front. Hardly before the sound of the shot died away, adrenalin coursed through his body and shocked him into momentary sobriety.

Shaken, he cried out in terror to the Virgin, "Holy Mother, the hand of God is upon us!" and sank to his knees, not so much in supplication as because his legs betrayed him.

And then, mortal fear subsiding, he arose and shakily made his way to the hills beyond town to ponder upon his punishment.

Little is left of the old Cannery Row; most of what has survived the ravages of fire is doomed by wrecker's hammers and the designs of the developers.

In the best tradition of such tales, it would be good to say that Pilon went forth and sinned no more. Alas! it was not so. His peccadilloes continued without noticeable pause until he died, not long ago, full of years and cirrhosis. But whenever he had need to pass that way again — beside the cemetery and the gully behind it — his step quickened and his look grew apprehensive until he had safely put this place of confrontation behind him.

During Monterey's history the town has changed. In its earliest days it was a collection of shacks made not of adobe but of logs brought from the nearby forest, chinked with mud, and topped by thatch. They clustered close around the church, limited by the space inside the surrounding stockade.

For the first few years the people huddled there, simply holding on, because they had no choice. To the Crown they were there merely to provide a Spanish presence against encroachment by Russian, English, French, and American interests, all of whom had territorial ambitions where this coast was concerned. The *montereños* were not skilled in much of anything except horsemanship, intrigue, and the intricacies of the *fandango*. They did not plant nor did they harvest; many of them subsisted only because friendly Indians brought them occasional gifts of food — a rabbit, a bit of seal meat, or a handful of some edible native plant. They often went hungry when the supply ship from San Blas or Loreto did not appear at its appointed time, which was the case more often than not.

The arrival of the supply ship, laden with provisions sent by the government in dis-

Windows where sequestered ladies of La Ida's menage once surveyed prowling males now look out on lines of automobiles and crowds of tourists.

tant Mexico, was a matter of great moment in the life of such a lonely outpost. It brought not only food and sundries but also news of the outside world. It was customarily a time of great rejoicing, and the townspeople entertained the ship's crew with hospitality. Fiestas and fandangos were the order of the day as long as the ship remained in port. Since it was customary that a suitable welcome be given as soon as the ship dropped anchor, the corporal of the guard was instructed to fire a round from the garrison's only cannon to mark the occasion.

One time, early in the history of the colony, the supply ship was long overdue; weeks stretched into months with no sighting of sail by the lookout on the hill. Gradually, the settlement was faced with actual starvation. Food in the storehouse grew short; weevils appeared in the flour; other stores grew moldy. Since there was little farming going on, the plight of the town grew desperate.

Some talked of deserting the presidio to seek help in some other place. Wiser heads pointed out that there was nowhere else to go; the nearest settlement, San Diego, lay five hundred miles away, with no assurance that it was any better off than Monterey.

It was in these depths of despair that the lookout on the hill looked up to behold great white sails against the blue of the sky beyond the Point of the Sea Wolves. A runner was sent to take the tidings to the commander, who caused the bells of the church to be sounded so that men idling beyond the walls of the stockade would receive word of the ship's coming.

They came running, their wives dashed out of the houses, children were gathered up from their play in the dusty plaza, and all the town ran headlong for the beach, where the ship could be seen already rounding the Point of Pines to the northwest — all, that is, except the corporal of the guard, who stayed behind with his squad to fire the ancient cannon, Spain's only piece of artillery in this bastion of empire, in a mighty salute of welcome as soon as the ship's anchor dropped into the waters of the bay.

The moment arrived when the ship hove to in the calm waters off the beach near Serra's great oak. The people on shore searched for familiar faces at the rail and in the rigging, where the sails were being lowered and furled. At the captain's command, the capstan was loosed, and the great chain of the anchor began to pay out, the sound of its going echoing against the hills around the town.

The corporal of the guard heard this sound and, mindful of the great responsibility resting upon him, advanced on the rusty cannon, which had been loaded with a generous serving of black powder, stuffed with wadding, and the whole charge tamped down with deliberate strokes of the ramrod. He stood as tall as his short form permitted, called his ragged detail to attention as befitted the magnitude of the occasion, and struck flint and steel together to light the fuse that led to the charge in the cannon.

On the first stroke he was successful, the fuse sizzled properly, and with a satisfying detonation the cannon went off, sending a great puff of white smoke across the parade ground of the presidio and startling gulls a mile away into circles shrill with cries of alarm. The corporal was in the process of congratulating himself for a duty well performed and anticipating the plaudits of his captain when the smoke thinned to reveal the flaming arc of the burning wadding from the cannon's discharge reaching the apogee of its trajectory and turning downward into the cluster of thatched roofs huddled between the church and the barracks.

In the confusion that followed, flames leaped from one torch-dry roof to another, consumed most of the houses of the forlorn colonists, leveled the barracks, and damaged the mission church severely. The records of history reveal nothing of the fate of the corporal who set this chain of events in motion.

Perhaps he did the people of Monterey a favor. The town was rebuilt; houses of adobe instead of logs began to be erected,

Where sardines once flowed from the sea into the channels of trade, the old canneries stand empty and silent, awaiting their end.

The canneries' wooden frames have burned away; only their concrete piers remain, like the ruins of some Grecian temple beside the sea.

and the more adventurous souls began to build beyond the limits of the stockade. Crowding and congestion were lessened. In time the village spread over much of the flatland of the present townsite. Toward the end of the last century, someone surveyed the town and laid out streets so that those who went from one place to another no longer trod the paths that cows had cut in the earth. A streetcar line was also laid out, from Monterey to the campground in Pacific Grove. The streets are still there; the street cars are long departed.

There are other changes. During World War II two military installations were established at the south end of the bay. Troops poured in along with the attendant camp followers: wives, children, and those hustlers of both sexes who have, since man first began to throw stones at his fellows, grazed on the largesse of military spending.

The limits of the town have crept ever more distant from the old presidio. That venerable establishment has long since disappeared; all that remains is the chapel, now restored to become Monterey's Catholic church. Where supplicants once beseeched the Virgin in soft, sibilant Spanish, voices now are more likely raised in the Sicilian dialects of Palermo and Isola delle Femine.

When the war ended, the military stayed on. Where before there had been only a sleepy cavalry post, three bustling military installations were firmly established. Fort Ord stood on the dunes along the shore; the old Hotel Del Monte had become the Naval Postgraduate School, and the Presidio the seat of the Defense Language Institute. The installations exist to this day in varying degrees of activity, depending upon the international exigencies of the moment.

The sardines vanished — fished out. The boats that gathered them were sold. The old cannery buildings went up in smoke and flame, and some hold to this day that the

Loving hands of modern owners have saved some old adobes, and today they are better cared for than they ever were by their original owners.

canneries' last planned harvest was the insurance money paid to their owners. But the tourists came in ever-increasing numbers, and Monterey and its neighboring cities woke up to the realization that too many old adobes had been leveled, torn down in the name of dubious progress. Those that were left became sacrosanct. They are looked upon with great affection by most *montereños*, both those who respect them as links to a historic past and those who recognize in them their real worth — Monterey's most effective magnet for the well-heeled tourist.

In recent years there has been a new change in old Monterey. Urban renewal tore down the old wooden buildings stretching down Alvarado and Pacific and Tyler to the wharf. In their place have risen many-storied hotels, a congeries of restaurants and gift shops of one sort or another, and a sprawl of multilevel parking garages.

Where there once was a kind of shabby antiquity with a *fin-du-siècle* air now stands an architectural melange neither Mexican nor American. In time, its detractors will come to grudgingly accept it; their grandchildren, never having seen the place as it was, will never notice the difference. But there is still a vocal contingent among the citizens who long for the lowly Mexican corporal to return once more and turn his cannon upon this modern reincarnation of Old Monterey.

Some of Monterey's old buildings are graced by famous names from the Civil War — Sherman, Halleck, Fremont, and more; others once housed the emporiums of New England traders, here to reap a profit from those who passed this way enroute to the Mother Lode.

106

PACIFIC GROVE

The town lying at the tip of the peninsula shares with Monterey just to the east an air of time past — though of lesser degree. Pacific Grove is hardly more than a century old, yet its older parts still bear the appearance of Victorian times. The houses do not have the distinctive western and Mexican flavor of those in Monterey; rather, they are more of New England and the Eastern seaboard — hardly surprising since most of the men who built them came from the other rim of the continent.

Pacific Grove had its beginnings in 1875, when a group of elder statesmen of the Methodist persuasion met in San Francisco and formed the Pacific Grove Retreat Association. Their purpose was to establish a "Christian Seaside Resort" on the shore of Monterey Bay, inspired by a counterpart established a decade before at Ocean Grove, in New Jersey. Even the town's name echoed the earlier one.

They bought a part of *El Rancho de la Punta de Pinos y Pescadero* from a Southern Pacific subsidiary and another parcel, to be used as a camp meeting ground, from David Jacks, a canny Scots Presbyterian who sealed the bargain by charging the divines one dollar for his land. They signed notes for the money due the railroad and laid out the parcel in small plots, each calculated to accommodate a tent during the summer when members of the church came to play and pray beneath the pines.

In the space allotted for the camp meetings, they raised a large tent and began to preach the Word every day all summer long. A chronicler of that time says there were services in the morning, services in the afternoon, services at night, and special services whenever an occasion could be found. Most services were long; some were short, but only a few. Everyone attended; no one was excused. Sinners were prayed over — at least, those who would admit to this condition — and some were momentarily saved. The woods rang with joyful hosannas. Large numbers of boys and girls and adults joined the Methodist Church.

They built a fence around the settlement, for reasons not fully clear a century later. Probably the fence was as much to corral the saved as to protect them from any touch of debauchery that wandering infidels from the fleshpots of Monterey might be tempted to introduce into these hallowed grounds.

In Pacific Grove some of the tents of the original religious encampment were converted into houses; a century later they still stand, small homes along the old streets of town.

The old lighthouse that stands among the wind-bent cypress at Point Pinos has been shining for a century and a quarter, only now it operates on electricity instead of whale oil.

There had been a house or two standing here even before the inception of this Christian Seaside Resort, and it was inevitable that those who came to live in tents and worship would soon long for more solid creature comforts. Small houses began to rise on the plots, many of them constructed by merely adding siding and shingles to the tent frames that already stood. Some owners simply enclosed their tents — the canvas became a lining for the walls and roof they built to keep out the chill of summer fog and the drip of winter rain.

The board of directors of the Pacific Improvement Company had as its members those same sharpshooters — Leland Stanford, Collis P. Huntington, Mark Hopkins, and Charles Crocker — who had built the Central Pacific Railroad, now become the Southern Pacific. These gentlemen were not inclined to give anything away. They had made a deal that while the churchmen should control the daily management and morality of this Christian Seaside Resort, their company would keep tight control of its financial operations. Seeing the rising tide of interest in the colony under the pines, the company added additional land to the original plot, divided this land into building lots, and began to sell it off. Im-

Some old buildings that once were homes have since been put to commercial use, but this one has been an inn ever since it was built a century ago.

mediately the peninsula's first real estate boom took form. Lots increased in value, and there was brisk bidding among both home seekers and speculators.

It was about this time that one Judge Langford, who had a summer cottage within the retreat, drove up to the fence one night, found it locked, pulled an axe from beneath the seat of his surrey, and neatly demolished the gate. It was never closed again.

Its demise seemed to increase the flood of eager speculators that swarmed in to bid the properties up. The small newspaper of that day exulted that real estate dealings amounted to $30,000 during one six-month period. The Retreaters, as the churchmen and their followers had come to be known, were not profoundly pleased by this state of affairs. Their protests were plaintive: "This is a Christian Seaside Retreat and not a place for money making." The money men in the San Francisco board rooms did not quite see it this way, and the beginnings were laid for a tug-of-war between God and Mammon that still goes on in more subtle form today.

The rules by which the community lived were set by the church hierarchy, who were strict in the regulations they promulgated to

insure the purity of those within this circle of the redeemed. No alcohol nor tobacco could be imported, sold, nor used; no vulgar language nor indecencies were tolerated; no clothing of less than decorous nature was tolerated; where there were windows, no shades could be lowered before 10 P.M.; on Sundays there could be neither work nor play. Particularly was it stressed that there would be no bathing without costume.

Opening the gates to all comers had admitted some decidedly not of the pious. The records tell of shifty types who wandered among the tents at night to watch the wondrous silhouettes that sometimes graced the canvas walls after the kerosene lamps inside were lit. There were others who drank of the grape and had to be rounded up and dealt with harshly by the law.

Liquor was always a problem until modern times. At the end of World War II statistics were quoted to show that there were more guzzlers, open and secret, in Pacific Grove than in any of the other towns. Whether this was true or not can never be sure — certainly it sometimes seemed so; what is known is that an emporium on a corner where Monterey and Pacific Grove met sold more spirits than any other in Monterey. Their customers must have come from Pacific Grove; nowhere else nearby was there such an unsupplied clientele.

The teetotalers and the church people fought off booze with local liquor laws; by the middle of the twentieth century, it was the last bone-dry community in California. But the balance had slowly shifted, and not long ago the diehards were overwhelmed at the polls when the voters removed the law from the books. Now a monumental liquor store stands in downtown Pacific Grove directly across the street from an empty lot where the huge old Methodist Church used to stand. Whether this exchange was for the best or not is debatable; there are still those who turn their heads away as they pass by.

The old town is not so different in spite of changes that time has gently wrought. Its streets, once grassy and immaculate, still sparkle on those days when fog withdraws and the sun beats down brightly. There is an underground population of the young who live in the small old houses of the Christian Seaside Resort and who sample frequently the heady smoke of *cannabis sativa;* there are hundreds of others who have never tasted anything stronger than frozen orange juice. Churches are still scattered all over Pacific Grove, and on Sundays the burghers and their families still come forth as their fathers did to pay their respects to their God.

It has more civic pride than any of the other towns; its waterfront is a spotless, pleasant place of trees and grass and lovely expanses of carpetweed which turn the tops of the sea cliffs into a blaze of lavender in the springtime.

Where Monterey has sacrificed too many of its ancient monuments for the sake of dubious progress, Pacific Grove has enhanced and preserved its own. In the bargain it has tempered its modern facade with a patina of age that has left it looking like exactly what it is — a lively, old-fashioned town, a jewel in the circlet of peninsula cities.

Most old houses in Pacific Grove reflect the influence of Midwest America rather than the Mexico whose legacy is seen in Monterey.

PEBBLE BEACH

The low range of tree-covered hills that extends west from Mount Toro in the Sierra de Salinas, bisects the Monterey Peninsula before it subsides into the sea between the two bays of Monterey and Carmel. On its southern and western slopes lie the lands that the Spanish once named *Los Ranchos de la Punta de Pinos y El Pescador.*

They were not long in the hands of the Barettos and Abregos, the original grantees; David Jacks soon had them firmly in his grasp, part of the more than sixty thousand acres he owned in Monterey County. Then, in the 1880s, the Big Four — Huntington, Hopkins, Stanford, and Crocker — bought this land and another parcel near the waterfront in Monterey from Jacks at forty dollars an acre, a tidy sum at that time and one that gave a fat profit to Jacks.

Their purpose was to build a great resort hotel in Monterey to attract the very rich who rode the rails in private cars and frequented such carvanserais of the elite. The land beyond the hills to the west was to supply the hotel with water from springs supposedly there in plenty. The water proved, however, to be in short supply, and

Crocker always felt that he had been robbed — a situation that most citizens of California savored greatly since the shoe was commonly on the other foot.

The Del Monte Hotel opened its doors in parklands as beautiful as any in America and was an instant success. Millionaires — those that Stevenson called "vulgarians of the Big Bonanza" — flocked there in droves; daily the trains from San Francisco brought more. It was said that on an ordinary day as many as five hundred trunks and other pieces of heavy luggage were unloaded from the train for transport to the hotel.

The lands along the ocean front and Carmel Bay lay largely empty during those early years. Stillwater Cove harbored a Chinese fishing village whose population varied strangely. One day there might be as few as twenty residents, only to have their numbers swell to several hundred on the next; then mysteriously on the third sunrise only the original twenty remained. It was probably no coincidence that Crocker, the construction genius of the Central Pacific Railroad, used thousands of illegal Chinese coolies in laying track through the High

The shifting palette of the sun on the Pacific could never be matched on any artist's canvas.

114

Sierra and across the Nevada desert. More than likely most of them first set foot upon American soil on dark nights in Stillwater Cove.

On a hill overlooking Pescadero Point stood the only house of those early years — the great log cabin erected as a summer home by the son of another buccaneer railroad magnate, James J. Hill of the Great Northern; a descendant of his lives there still.

These lands at the western tip of the peninsula were laid out with riding trails and a scenic road which hotel guests traversed in great tally-hos pulled by three matched pairs of lively horses. The road led from the hotel through Monterey and Pacific Grove into the dunes and forest along the ocean-facing cliffs. Eventually it wound around and back over the mountain to return to the hotel. The brisk ride required several hours and covered seventeen miles from beginning to end. The descriptive name given it has persisted for a century; the 17 Mile Drive is probably as well known as any road in California, and better than most.

A disastrous fire leveled the sumptuous hotel before the turn of the century, but it was rebuilt, more ornate and spacious than before. Beyond the hills to the west the land lay unused except by horseback riders cantering along the paths, sightseers in the carriages that circled the drive, and picnickers from the hotel dallying on the dunes and the mounds where once Indians dwelt among the pines.

No Chinese remained here by the end of the century. They disappeared almost overnight, and their shacks were leveled. Historians of our time know nothing of their fate; no one at that time cared enough to mount a persistent inquiry into their vanishing.

By the end of the first decade of the twentieth century, the big hotel had declined as a resort, becoming stodgy and dull under management grown too old and overly pious at the helm. Guests chafed under social restraints many found petty and some downright suffocating.

The moguls of the Pacific Improvement Company did not take kindly to the prospect of having a first-class white elephant on their hands; besides, agents of the government, intent on enforcing provisions of the newly enacted Sherman Anti-Trust Law, were beginning to snap at their heels. It seemed as good a time as any to summon S. F. B. Morse from the San Joaquin Valley, where he had been overseeing the administration of the company's vast land holdings, which stretched from Seattle to New Orleans. He was instructed to pay special attention to the Monterey Peninsula operation and to liquidate it as soon as possible.

Sam Morse came, looked closely, and saw the opportunity that others had passed over. He saw the empty land that lay along the famous drive and recognized a real estate bonanza of great proportions lying fallow there. He brought in new managers to revitalize the hotel and began to look about for a way to gain control. He went to a New York bank to borrow a million dollars, and took a San Francisco financier and associates as partners for another million. They bought the land, formed the Del Monte Properties Company, and sold stock

Along the dunes on 17 Mile Drive, winter rains leave pools that reflect the brilliance of winter sunsets.

The pine-covered hills of Pebble Beach, and its golf course, abut the town of Carmel; both share the mile-long beach of sparkling sand, which has no superior along the California shore.

Meticulously manicured greens of golf courses lie in stark contrast to the untamed growth of the cypress that surround them.

for a third million to gain operating capital.

At this point Morse took a step that was to be an example to others on the peninsula during the years that followed. He found that the land stretching along the cliffs from Carmel Beach to Pescadero Point had been divided into small lots intended to be sold at prices far too low for such prime land. He tore up the subdivision map and declared this strip a green belt, dedicated in perpetuity as open space. To keep it that way, he set out to turn it into a golf course, which in time became one of the most famous in the world.

At the end of the golf course he built the Del Monte Lodge and when the old Del Monte Hotel in Monterey burned down again, he rebuilt it of stone and brick and mortar instead of wood. Morse had no intention of maintaining these two plush hotels merely as a service to the extremely wealthy; instead he intended to use them as magnets to attract rich patrons to buy his real estate. To sell it, he hired men of stature as sportsmen and women of social distinction. They wined and dined the guests in a heady whirl of dancing and champagne and romance. More than one deal for choice land was consummated in the small hours of the morning.

Great houses with many rooms and spacious grounds began to rise among the old live oaks along the edges of the fairways, each house standing well apart from its neighbors. Celebrities were a commonplace in those days, as they are today. One man of mature years tells of a childhood in which he came down to breakfast to find Douglas Fairbanks at his father's table one morning, and Charlie Chaplin and his current love there the next.

Morse was the corporation's overlord for half a century — a benevolent despot. He brooked no opposition and heeded little counsel save his own. But when he died, he left behind him an intact forest laced with golf courses known around the world: Pebble Beach, Cypress Point, and Spyglass — a green and lovely park that will be his monument as long as those who have succeeded him hold to his vision as their blueprint for the future.

CARMEL

The road south from Monterey climbs over the ridge that forms the backbone of the peninsula and descends toward the mouth of the Carmel Valley. At the top and on the Monterey slope, the road is a freeway now. A traffic interchange that is sometimes simple cloverleaf and sometimes puzzling maze sits atop the crest of the hill. When bulldozers first appeared to level the forest for this huge traffic artery, one bright-eyed Carmel lady who was given to espousing lost causes attempted to prostrate herself in the path of the roaring monsters in order to stem the tide of change that threatened to engulf her world. She was led away by huge deputies from the sheriff's office, who smarted under the tongue-lashing the gentle lady unleashed upon them.

The road was built — it was desperately needed, and time has proven the fact. Most people have forgotten the furor that attended its proposal, and the pines and cypress that have grown thickly there in the intervening years have covered the dreadful scar and stilled the protests.

But on the Carmel side, the road is as it has been since the days when cars went more slowly up the steep hill and there were fewer of them. Now the volume of traffic sometimes chokes its narrow width, and lines of stalled cars a mile long are not unusual enough to provoke serious comment.

The reasons for this are not particularly complex, nor are they puzzling to those who have known Carmel for even a short time. The town simply balks at conforming to what other California communities consider right and proper. It has used its not inconsiderable influence to sidetrack the efforts of the state's concrete and highway interests to complete the raceway that ends in a bottleneck of pure frustration miles short of its intended terminus.

This same attitude has worked for years to forestall sidewalks and streetlights in the residential areas, numbers on its houses, neon lights in the business district, and mail delivery to anyone within the city limits. It came near at one time to preventing the paving of any of the town's streets; only after vehicular traffic grew so heavy and potholes of such awesome proportions that Ocean Avenue gained the name of "Devil's Staircase" did the city fathers take steps to provide this amenity.

Such resistance has worked well enough

In early morning the streets of Carmel are empty — a short-lived pause before daily crowds come to inundate the town.

in the three-quarters of a century the town has been here, well enough that a majority of its residents want no particular change. It has made the town attractive to new residents and visitors from all over the world, and the price of its real estate has gone up almost beyond the limits of credibility. Houses that in other places might go begging for a buyer sell here for half a million dollars and more; no one seems to think this particularly unusual.

It was not always this way. The first efforts to capitalize on the undeveloped charms of the south slopes of the mountain came in 1888, when a citizen of Monterey named Duckworth looked upon the success attendant to the establishment of Pacific Grove as a retreat for devout Methodists and decided he could do equally as well with Roman Catholics in the shadow of the Carmel mission.

He was wrong. He laid out a grid of streets — on paper, at least — on the high slope and let the word be spread that these admirable lots were available for a mere twenty dollars each, a steal for anything so

At dusk the lights of Carmel and Pebble Beach shine through the pines and onto the dark water, while on the distant hills the lights of Santa Cruz twinkle like tiny stars.

close to the sparkling beach and the romantic old church. He even hinted that the Southern Pacific would soon extend its tracks to Carmel City, a boon to weary travelers who would no longer have to ride the stage from Monterey and be obliged to dismount and walk the steepest part of the hill to spare the straining horses.

Duckworth was the first of a long line of real estate promoters, but unlike the restraints upon his modern counterparts, the law was more relaxed and he was able to traffic in the truth stretched a little beyond its normal limits in promoting his prospective Eden. There were no trees on the slope that stretched down to the river and the bay, the nearest store was five miles away in Monterey, and there was no water. The walk to the mission was two long miles through brushy gullies and across dunes. And the Southern Pacific Company apparently was not privy to Duckworth's prospectus; when the line into Monterey was abandoned eighty-odd years later, track still had not been laid to Carmel City.

The hoped-for throngs of Roman

Catholics apparently found other interests to absorb their attention. A few lots were sold but there was no overwhelming rush to build on them. There was nothing on that hillside to impart status — a thing important to Americans of that time, a thing important to Americans still.

Not until James Devendorf came upon the scene was the idea of a town revived. He traded land he owned in Stockton for Duckworth's holdings and other parts of the slope closer to the sea. His partner was a San Francisco attorney, Frank Powers. Like Duckworth, they imposed a grid pattern on the irregular Carmel terrain. On paper it looked good, but on the ground it was another thing; where there were gullies, there was the inconvenience of dead-end streets — streets that in today's Carmel are cul-de-sacs of peace and quiet.

Their idea was a community of cultural bent — a colony of artistic types that would appeal to professors from Stanford and Berkeley who had money to spend and would be likely prospects. To attract the sort of customers they sought, Powers and Devendorf needed a lure; they got one in George Sterling, a well-connected Piedmont socialite, a poet, and a man of wide acquaintance in art and literary circles. They made a deal with him — probably a plot of land in return for a promise to build — and he came down in 1905. Mary Austin and Arnold Genthe quickly followed, and the rush was on; it still goes on after seventy-five years.

The decade that followed was singular; the coterie that grew around the nucleus that was George Sterling reads like a Who's Who of literature and art in the West: Joaquin Miller and Charles Warren Stoddard, Jimmy Hopper and William Keith, Jesse Lynch Williams and William Merritt Chase, Harry Leon Wilson and William Rose Benét, and an uncounted host of others. Some of them stayed only for a while — Sinclair Lewis and Upton Sinclair made themselves so unwelcome that their time in Carmel was not prolonged, and Jack London came only twice on short visits, although some hip-pocket historians are quick to claim him resident. Others came and never left.

Robinson Jeffers and his new wife moved north from Los Angeles in 1914. They had planned to live in Ireland, but the sound of guns in Europe that autumn turned their path awry, and they made their way instead to Carmel. There, on Carmel Point, Jeffers built his house of stone, laboriously carried up from the beach nearby, and spent the rest of his life writing

Robinson Jeffers, the poet of Tor House, has been gone for more than twenty years, but the trees he planted and the stone towers he built are monuments to his being; the words he left behind make his name honored wherever the English language is spoken.

and contemplating the restless sea before him. He took the rugged coast of the Sur for his settings; the people of his pen who lived there have become an inseparable part of the literature of America.

Tor House and Hawk Tower still stand where he built them. The trees he planted around his house are there yet — stately cypress and tall eucalyptus swaying in the wind that brings the gray sea fog he loved so well. His ghost still climbs the steep stairs of the tower to look out over the ocean when distant storms send waves crashing against the sea cliffs at the foot of his wild garden.

There were others here in his time but none that was his equal; Mary Austin had left before Jeffers arrived. None of the others who lived in Carmel ever reached the lowest of the heights that Jeffers' poetry attained. Those who had talent were often content to waste their time in talk and disport; those without it reveled in the re-

flected glory of their more blest brethren.

Still, the Bohemia that was Carmel in the first two decades of the century was known far and wide, attracting poets and writers and painters like moths to a flame. Most of them, both with talent and without, are remembered less for their accomplishments here than for their participation in the heady life of that time.

Life was easy. Daisy Bostick, a cheery little soul who wrote of Carmel in those early days said: "The artist who comes to Carmel finds that he can secure a three-room bungalow, studio, or shack for an extremely moderate monthly rent. . . . Aside from his smokes and the material for his work, what else does he need that money could buy? . . . Today's ten cents is as good as

Artist Joan Savo and cartoonist Gus Arriola are known far beyond the limits of the peninsula — Savo for her paintings, Arriola for his comic strip, "Gordo."

Where once only this artist's-owned cooperative showed work of local artist, now dozens of private galleries — some excellent, some atrocious — crowd Carmel's streets and byways.

yesterday's dollar. Things that mattered so vitally in cities — bank accounts, conventional clothes, keeping up appearances — seem no longer of much importance."

Daisy would be aghast if she should come back today. Some of the very same bungalows, studios, and shacks that were here in Daisy's day are beyond the reach of any struggling artist, were he to seek to rent one of them; today's dollar being only as good as yesterday's dime, a month's rent might well be equal to the original cost of construction.

There are other things that have not changed as drastically. Some of the popula-tion wear conventional clothing; the rest do not. But the unconventional clothes may themselves be a uniform — today's rags and tatters more often than not bear a designer's label. There are still picnics at the beach — but no longer is there wine and abalone cooked over a driftwood fire; the new Bohemians tend more toward Coors and hamburgers sizzling on a portable hibachi.

Where once there was only the cooperative owned by the artists, there are now almost a hundred galleries. Where once there was on Ocean Avenue only Slevin's General Store and Post Office to dispense groceries and gasoline, postcards and patent medicines, stamps and sundries, there are now warrens of shops of extraordinary form and variety tucked away in arcades and courtyards filled with trees and shrubbery. Stores that used to exist to serve the local people can no longer afford skyrocketing rents. They have closed their doors and moved to Monterey or the sprawling complex at the mouth of Carmel Valley.

So a division that was inherent from the beginning has grown to threatening proportions. There are now two Carmels — the town that exists to entice the tourist and extract his dollars as quickly as possible and the other town that curses the tourist and wishes he would begone forever. Nowhere are the battle lines more evident than in the chambers of government, where business interests are lined up against those who purport to represent the residents' interests. The two sides are increasingly polarized and, as time goes by, find less and less common philosophical ground.

These two towns occupy the same space,

Weaver Mary Buskirk is a major innovator in using fibers to create works of art; Ansel Adams has long been preeminent among modern photographers.

but the outsider sees only that circumscribed within the limits of the crowded business streets, the mission, and the white beach beside the bay. The other town is a hidden place of quiet gardens, music, books, and interests of such diversity as to encompass much of the range of human learning. This town teems with retired professors, former executives of commerce and government, and a generous sprinkling of immigrants from most of the civilized countries of the earth — and a few less so — a town largely of the middle class. There are a handful with great wealth and some who exist in a fringe life of food stamps and government subsistence checks.

In the beginning there were few trees on the slopes of the mountain where Carmel now lies. The original developers took note of this lack and planted a generous number of oaks and pines. Most of them now approach the century mark and are the giants of the urban forest that covers the entire town.

The windows of houses along the sea catch fire at day's end as the sun takes one last look at the darkening land.

Many houses look like interlopers displacing what was there before them,
but a welcome few seem to be an inseparable part of the forest.

On weekends Carmel sidewalks become a solid mass of humanity that surges restlessly back and forth from one shop to another . . .

Carmel's houses, though crowded among these trees, are separated by fences and masses of shrubbery that provide great feelings of privacy — one of the town's major charms — so that people can withdraw into their own green and flowered enclaves and shut out the world's prying gaze. There is a feeling among the townspeople that a few have pulled their gates shut and have not come out in years.

Carmel takes pride in one thing that brings the whole town together and often: no postman walks the streets of Carmel. Since there are no numbers on the houses,

no mail is delivered. Each person who lives within the limits of the city must either go to the post office or give up written contact with the world outside. There have been moves afoot from time to time to change this arrangement, but every attempt is shouted down in anger by a vocal majority. Most who live here look forward to the daily pilgrimage to the post office; indeed, days when the post office is closed are often aimless ones for a sizeable portion of the population — those who live alone, the aged, the widowed. Rituals such as this are not easily surrendered, even if only for a day.

The people of Carmel still turn out in numbers when a flush of color in late afternoon presages a sunset of even average intensity; spectacular ones bring out half the town. The more settled citizens stroll along the cliffs or walk the beach just beyond the highest reach of the waves; the younger ones mostly flock together in parked cars above the spot where black-clad surfers ride the measured pulse of the tide and the heady breezes of marijuana mix gently with the winds from the sea.

On such days the sun drops into the sea with an almost audible splash; dusk creeps through the pine-lined streets and paves the way for the coming of the dark. Lights shining through half-shuttered windows cast a warm glow and silhouette the twisted oaks that walk the night in Carmel gardens. The smoke of burning pine logs drifts upward from a hundred chimneys and creeps through the forest — a fragrant blue haze in the twilight. It is a lovely time. Those who live in this place think it near perfect.

. . . but on the beach there is always room to spare.

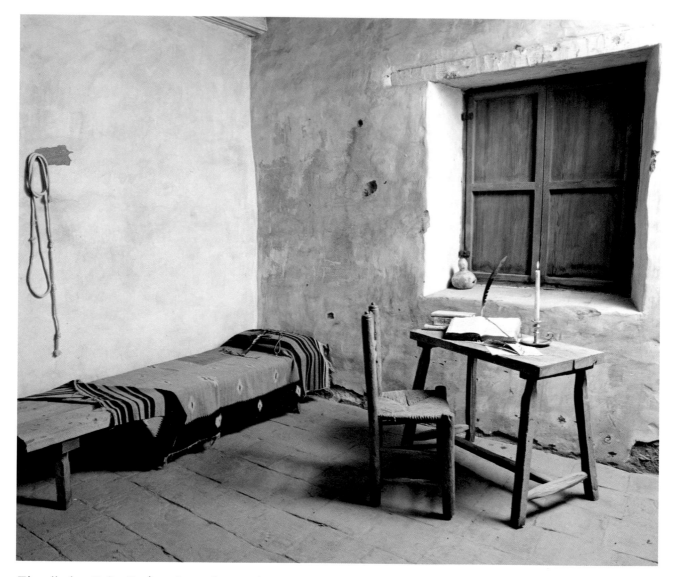

The cell where Padre Junípero Serra slept, worked, and died is Spartan in its appointments, but probably as luxurious as this saintly man ever saw.

The first light of day catches Serra's marble monument and finds his friends Crespi, Lasuen, and Lopez mourning at his bier.

On special occasions luminarios *light the paths through the mission garden, where thousands of Indians lie buried.*

Two centuries have gone by since Serra put his artisans and Indians to work raising the first stones of San Carlos Borromeo.

*Only a year or two ago the house still stood where friendly ranchers had taken
Robert Louis Stevenson to rest when they found him stricken on a mountain trail
above Carmel; now time, weather, and heedless vandals have brought its ruin.*

*When Mexico secularized the missions, their lands were passed out to cronies and retainers of those
in power in Monterey; the adobes those settlers built still stand after a century and a half.*

Ocean fogs creep into the mouth of Carmel Valley; beyond, the sun shines brightly on the valley's hills.

BIG SUR

For over a century and a half after the authority of Spain was established in Monterey, the coast to the south remained a place of extreme remoteness. Roadless, it was a haven for those wild things that move on when man moves in. No one came into these coastal mountains in the early years unless he was lost or a fugitive from the indifferent arm of Spanish law.

No man held title to any of the land for almost sixty years, until Governor Jose Figueroa bestowed *Rancho El Sur* on Juan Bautista Alvarado and *Rancho San Jose y Sur Chiquito* became the domain of Teodoro Gonzales. Neither of these grantees was possessed of an iota of pioneering spirit; they were political connivers pure and simple, and the lands were given to them as a reward for loyalty to the establishment. It may be entirely possible that neither Alvarado nor Gonzales ever visited his lands.

Rancho El Sur, in time, passed from the lax, improvident hand of Alvarado into the possession of his hard-driving and energetic uncle, Juan Bautista Roger Cooper, an Englishman who had married wisely into the land-holding aristocracy of this Mexican outpost. Cooper ran cattle on the land

and built a house there, probably the first in Big Sur.

More houses were built to shelter *vaqueros* who worked the ranch. In the valley beyond and along the coast, other pioneers came and claimed land for themselves — Post and Pfeiffer, Partington and Dani and Harlan. These and others left their names upon the land. Some of their descendants still live where their grandfathers and great-grandfathers first staked out claims.

In those early days, life was too rugged to accommodate any except the most determined, and newcomers were slow in coming. Those who came for the most part were content simply to remain there. Some who lived in Big Sur in those times had been to Monterey only once in their lives; there may have been others who never got that far.

There were a few who made their way into the wildness of the Sur Coast during those earlier years, but only a few. Men did not come here on a casual whim. Enormous will and effort were required to surmount obstacles that nature had put in the path of those who made their way here.

Probably no one ever had as difficult a

Along the Big Sur coast, rugged mountains drop directly into the sea and waves cut away at the overhanging cliffs.

time as Harrydick Ross and his wife, when they grew impatient with life in Los Angeles sixty years ago and set out on foot to find a new home. They walked more than three hundred miles and scaled the mountains that separated the ocean from the interior valleys. They found a place on the shoulder of Partington Ridge high above the sea, and there they built a home. His "Shannah Golden" is gone but Ross lives on there alone, a patriarch beloved by his neighbors.

Change was slow in overtaking the south coast because entry was too difficult in this place of steep, eroded trails and deep wagon ruts. So during the first century after his arrival, western man had remarkably small effect on this wild place. But his insistence that all things be made easily accessible finally brought the outside world to Big Sur. In 1937, convict labor completed the last link in the paved road that skirted the edge of the sea from San Luis Obispo to Monterey.

The highway brought tourists, but at first only a few. More settlers arrived, too, but strictures of war prevented any major influx of people. The country was preoccupied with more pressing matters, so for the better part of a decade Big Sur was left to itself.

World War II had hardly ended when Henry Miller appeared one day on the doorstep of his old friend Emil White at Anderson Canyon. He lived in those times on the edge of penury, denied publication of his books by censors in America, and survived on whatever royalties his European publishers sent him. In spite of this he lived a life that some called hedonistic, and others described in more accusatory terms. He ignored them all and lived as he pleased.

He wrote of the good life in Big Sur, attracting an avalanche of beats and hippies to partake of the carefree bliss of this coast. At almost the same time the Hearst family gave their father's palace to the state to avoid its enormous tax burden, and the state promptly made it into a historical monument that proved to be one of the most enticing tourist traps in America. Hordes of visitors drove down the coast to reach the castle, a symbol of what most Americans put great store by — immense wealth and the outer trappings that proclaim it. It was all too much for Miller; he fled to spend the rest of his days in Los Angeles — a most unlikely haven.

Hotels and motels, restaurants and gas stations sprang up in profusion to reap the beneficence of this endless harvest of tourist dollars. Houses proliferated along the ninety-mile stretch of highway as land was subdivided and sold off to land seekers. Those of lesser means built modestly beside the road; those of greater resource, largely on hilltops and higher slopes, often unseen by travelers below. Yet, all too often, bulldozer blades have left unhealed wounds on the slopes that reveal where these new despoilers are hidden.

Some newcomers see no despoliation, only a haven of peace in an unfriendly world; others see a land of opportunity ripe for the picking. Such is the trouble with latecomers; their frame of reference is such that "how it was" goes no further than "how I first saw it." Those of longer years in this place possess a different set of memories.

Margaret Owings, protector of sea otter
and mountain lion, lives in a house poised high
on a cliff above the ocean.

Sculptor Harrydick Ross, who first came to
Big Sur sixty years ago, works amid the clutter
of a thousand projects in progress.

Still they come. Every canyon, every hilltop holds someone's hideaway. There are artists and craftsmen, writers and musicians scattered along the coast; some are of real talent, some of none. Stellar lights of the entertainment world live cheek by jowl with remittance men and retired stockbrokers. The cost of subsistence is not cheap, and an excursion to the supermarket in Carmel is a long seventy-five-mile round trip. To live in Big Sur nowadays one must have either adequate funds or the brass to live on the backs of those who have.

There are residents who would call a halt — complete and final — to further incursions on the land along this ocean front. Some who hold this view have been here for four generations or more — since the time when their ancestors usurped the land of the peaceable Indians who lived on these slopes; others are new upon the scene. Those of opposing view, who look to profits to be made from their holdings, cry loudly against interference from outsiders. Some of these, too, are long upon the land.

As these words are written, there are new proposals to transform the entire coast into a national holding to preserve it. Those who live here generally oppose this threat to their autonomy, and partisan battle lines of different view are forming. Suspicion and distrust are turning neighbor against neighbor, friend against friend, and veiled threats with sinister overtones are not unknown.

These mountains have stood steeply out of the sea for unnumbered millions of years. In all that time the elemental forces of nature have shaped them into the marvelous form and texture they possess today. Yet in less than two centuries man has made marks upon them that may have results as profound as any made in the long millennia before.

Whether the land remains in private hands or becomes a national park is of little moment. In either case, man will be here in numbers and therein lies the danger. Man can rape the land and so despoil it; his love for it can often have the same effect.

*On the brink of a precipice a thousand feet above the cold, restless Pacific,
friends share a glass of wine in a bubbling hot tub.*

River and highway wind their way together from the high ranges down the Big Sur Valley toward the sea.

High upon a mountain above Pfeiffer Beach stands
a rusty Don Quixote, guarding the same land
his contemporaries first looked upon four centuries ago.

On summer afternoons, while inland valleys sizzle, residents and visitors alike bless the special blend
of fog and sun, sea and stone that is the Monterey Coast.

EPILOGUE

John Steinbeck once wrote something to the effect that he was angered when he returned to Monterey after a long absence and found the town changed. Then came a realization that the difficulty lay in the fact that while the town had changed, he had not changed with it.

For me there has been no absence. During the forty years that I have lived on the Monterey Peninsula, things have changed — and I with them. In far too many instances, the changes wrought in both brought less than improvement. But many have been so gradual that they have happened unnoticed until now I stand and look about me and realize that something is missing, something is gone, and I can remember neither when nor how.

Big events are well remembered: few who were there that day can forget when the wrecking ball of the huge crane drawn up on Ocean Avenue slammed for the first time into the ornate white facade of the old Carmel Theatre; and few can help but recoil from that which took its place. The majestic old spired Methodist Church that towered over Pacific Grove — beloved link with the town's religious beginnings — has been gone for more than a decade, and in its place is a huge vacant lot that serves no discernible purpose. And downtown Monterey — half of it has been torn away and replaced by other buildings of dubious form and uncertain worth.

There was the night when the sleeping village was aroused by the wail of fire sirens, and the populace turned out almost en masse to watch the Golden Bough Playhouse go up in flames — as dramatic a production as the old theatre had ever presented. It was a signal occasion for Carmel. Some who were there that night were seen in public for the first time in years; some have not been seen since.

Cannery Row is nearly gone, burned down — some say for profit, some say for revenge — and what is left has been converted into a gathering place for tourists. Only a superficial resemblance to the original old street remains, and that may disappear; even now the last of the working canneries disappears under the wrecker's crowbar.

The sense of loss may not stem wholly from past actuality; nostalgia induced by the uproarious pages of a Steinbeck novel has dimmed the memory of the stink that often floated from the canneries to blanket the whole peninsula.

With the going of the fish, the sardine fleet departed for more productive waters. The anchorage within the breakwater, which once was host to a hundred purse seiners dipping and curtsying in the swell, now holds only a few trollers and a swarm of pleasure craft crowding closely together in the new marina between the two old wharves.

Far down the coast, where the hot springs on Tom Slate's old farmstead bubbled forth and the overflow filled old bathtubs set upon the cliffs overhanging the sea, we went often on moonlit nights, tiptoeing past the innkeeper and down the slope to lie in parboiled bliss while the candles we had brought guttered out and only moonbeams and starshine danced on the sea surge below and bounced off bare skin just beyond visual reach. The tubs are concrete now, and we unbelievers are no longer welcome at the esoteric enclave that is Esalen.

Carmel, the old town that made me welcome forty years ago when I came as a captive transient, has changed. I have noticed the change and not noticed, not wanting to notice until in recent days the changes could no longer be ignored — downtown Carmel abandoned to tourists, its citizens crowded from their streets and forced to the mouth of the valley or to Monterey to trade, pre-empted by a flood of strangers in cars that jam the streets or in tour buses that befoul the air, a horde welcomed joyfully by the chapmen who keep the shops, but detested by the rest.

When artists gather now, one seldom hears the heady philosophical talk of aesthetics; instead, they talk of selling. The marketplace that is Carmel has attracted hacks as honey does flies, and the more serious artists have either fled or show their work here rarely if at all. Too few of talent and dedication remain; the rest follow formulas that guarantee financial return.

Of course, I am biased by the passage of years and the memory of a different time and style. Those who arrived only yesterday can have no such knowledge and no doubt will in time look back upon this moment as "the good old days."

There are still things here that transcend all changes that man has wrought upon the place. The white beaches of Carmel and Asilomar and Monterey are still a solace to those who love the feel of hard, wet sand with cold waves curling around bare feet or the light as a declining sun slowly changes the world from blue and white and green to gold and red then all the muted violets of gathering dusk.

Tall pines still answer to winds that bend their trunks in unison like a Greek chorus swaying on some sylvan stage. Cypress still stand lonely guard along the Lobos headlands, and beyond them pelicans course along in ponderous majesty, rising and falling with the rhythm of the sea — a fascination that will persist as long as such birds take wing along these rocks.

Waves still spend themselves upon lonely beaches at the bottom of sheer Sur cliffs. Giant redwoods stand close along creeks that have worn a cleft into the towering frontal range. Wild things that seek protection here still shyly reveal themselves to those who walk softly and quietly among them.

Old adobes that recall a forgotten time still stand amid the bustle and bluster that mark the old capital today, and old houses in Pacific Grove — both mansions and those more modest — take on new luster as loving owners find pride in these treasures of a century past.

Still — man comes, and in ever-increasing numbers. Slopes whose oaks have stood untouched for centuries hear the sound of chain saws and trees falling to make room for more and more houses. Each year the demands mount — demands for more water, more roads, more of all the services man has come to expect as his due, and loud demands for fewer restraints upon him.

Whether this coast can survive the pressures that man is putting upon it is a question whose answer is crystal clear to some, uncertain to others, and — to far too many — of no concern at all. There may be small reason to assume that man may rise above himself here — his record elsewhere has been less than an encouragement.

ACKNOWLEDGEMENTS

There is an enormous difference between photographing something not seen before and photographing that which one has grown accustomed to seeing — or rather, not seeing — over a period of several decades. And so this book presented me with problems in observing Monterey and Carmel, which I have so long traversed with my eyes shut to avoid seeing changes that disturb me. By comparison, Big Sur, the seacoast, and the mountains presented few such difficulties; Nature has a habit of doing things the right way, whereas man often leaves a heavy hand on everything he touches.

For better or worse, the task is done, the pictures made, the words written. Some things may not be seen as the outsider sees them, but there are other books on the subject for those who want nothing more than a visual tour; mine is an expression of how I *feel* about this beloved meeting place of sea and land.

Most books are created azygously. But before that time of solitude when pen is set to paper, there are usually others who help to prepare the way. This book certainly owes much to others who helped me immensely during its period of gestation.

Probably foremost were Emily Woudenberg and Peg Richter, reference librarians of Carmel's Harrison Memorial Library, who often went far beyond their normal duties in searching out long-forgotten details of the history of the Monterey coast.

Jean Grace of the same library was a veritable gold mine of information on the geography of the region, as well as a stimulating source of philosophical argument when it was most needed.

I am grateful to Lacy Williams Faia for her recollections of what it was like to have grown up in Carmel and for permission to quote from the book, *Carmel at Work and Play* by Daisy Bostick. Virginia Scardigli was a source of great knowledge concerning the Depression years in Monterey and on Cannery Row, as was Jimmie Costello in his recollections of his days as a young reporter on the *Monterey Peninsula Herald* when the peninsula was less crowded and more leisurely than now. Frank Lloyd was a literal encyclopedia of the old days in Carmel, and John Boit Morse was most helpful in revealing facets of his father heretofore unknown to me. My deep thanks go to all of these for sharing their memories and experiences with me.

On an extremely personal level, my everlasting gratitude to Rudy Propach, Bill Breneman, and Dave Erickson — friends who bolstered me and gave me hope through difficult and anxious days. Finally, there can be no way to fully express my abiding love and appreciation for my greatest admirer and severest critic of forty years' duration — Cookie Crouch. Without her, this book and much of whatever else I may have accomplished would have been infinitely more difficult.

NOTES ON THE PHOTOGRAPHS

Jacket photo: Sunset, Carmel River Beach.

Page 1 Waves among underpinnings of a ruined cannery, Monterey.

Pages 2-3 Beach at Granite Creek, Big Sur.

Page 4 Wharf and boats, Moss Landing.

Page 9 Monterey cypress *(Cupressus macrocarpa),* Pebble Beach.

Pages 10-11 Brandt's cormorants *(Phalacrocorax penicillatus),* Point Pinos, Pacific Grove.

Page 12 Surf and rock, Villa Creek Beach.

Pages 14-15 Cypress grove, Point Lobos.

Page 17 A sea urchin *(Strongylocentrotus purpuratus)* and a barnacle-covered sand dollar *(Dendraster excentricus),* Monterey Bay.

Page 19 top Sandpipers, Monterey. *bottom* Cormorants, Pacific Grove waterfront.

Page 21 Monterey pine *(Pinus radiata)* on dunes, Pebble Beach.

Page 23 Pine forest in fog, Point Lobos.

Pages 24-25 Monterey cypress in bluffs above Carmel Bay.

Page 27 Cypress and pine on sea bluffs, north shore, Point Lobos.

Page 29 Willet *(Catoptrophorus semipalmatus)* on rock ledge, south shore, Point Lobos.

Page 32 Common sea star *(Pisaster ochraceous)* in company of several bat stars *(Patiria miniata)* in tidepool, Weston Beach, Point Lobos.

Page 33 Giant green anemones *(Anthropleura xanthogrammica),* Weston Beach, Point Lobos.

Page 34 Sea palms *(Postelsia palmaeformis)* in surf, Point Lobos.

Page 35 Surf on rock ledges, Point Lobos.

Pages 36-37 Eroded sandstone rocks, Weston Beach, Point Lobos.

Page 38 Kelp floating in deep pools, Point Lobos.

Page 39 top Sea otter *(Enhydra lutris),* Point Lobos. *bottom* Cormorants on bird rock, south shore, Point Lobos.

Pages 40-41 Sea lions *(Zalophus californianus).*

Page 42 Beach at sunset, Bixby Creek, Big Sur.

Pages 44-45 Mouth of Bixby Creek, Big Sur.

Page 46 top Sea foam and kelp, Partington Point, Big Sur. *bottom* Early morning, Willow Creek Beach, south coast.

Page 49 Surf at Granite Creek, Big Sur.

Page 50 Fog at mouth of Little Sur River, Big Sur.

Pages 52-53 Sea bluff at Pacific Valley, south coast.

Page 54 Sea dunes near Point Sur.

Pages 56-57 Surf, Point Lobos.

Page 58 Succulents on dunes, Little Sur River Beach, Big Sur.

Page 59 Sycamores and road, Andrew Molera State Park, Big Sur.

Pages 60-61 Pampas grass *(Cortaderia selloana)* on sea bluffs, Limekiln Creek, south coast.

Page 62 Wild flowers and oaks, Carmel Valley.

Page 65 San Antonio River, Santa Lucia Range.

Page 66 Sun in Carmel during forest fire in Santa Lucia Range.

Page 67 Fire aftermath, Post Summit, Big Sur.

Page 68 Redwood trees, ferns, and bay trees, South Fork, Bixby Creek, Santa Lucia Range.

Page 69 top Western wake-robin *(Trillium ovatum).* *bottom* Manzanita and toadstool on dead tree, Santa Lucia Range.

Page 71 Fog at sunset, Plaskett Ridge, Big Sur.

Pages 72-73 Late-afternoon light on Plaskett Ridge, Santa Lucia Range.

Page 75 Winter on tributary creek, Carmel Valley.

Pages 76-77 Coast live oaks *(Quercus agrifolia).*

Page 79 Wind-twisted oak, Carmel Valley.

Page 80 California poppy *(Eschscholtzia californica),* creamcups *(Platystemon californicus),* and baby-blue-eyes *(Nemophila menziesii).*

Page 81 top Star lily *(Zigadenus fremontii).* *bottom* Fiesta flower *(Pholistoma auritum).*

Page 82-83 South Dolores Street, Carmel.

Page 85 Fourth of July, Carmel Beach. Because of extended drought, such fireworks are now banned.

Page 87 Fishing boats, Monterey Bay.

Page 88 Marina, Monterey.